WHAT THE FUTURE HOLDS

THE FUTURE OF ENERGY:

FROM SOLAR CELLS TO FLYING WIND FARMS

BY M. M. EBOCH

CONTENT CONSULTANT:

PETER C. BISHOP, PH.D., APF

TEACH THE FUTURE.ORG

HOUSTONFUTURES.ORG

CAPSTONE PRESS

a capstone imprint

Capstone Captivate is published by Capstone Press, an imprint of Capstone.
1710 Roe Crest Drive, North Mankato, Minnesota 56003
www.capstonepub.com

Library of Congress Cataloging-in-Publication Data is available on the Library of Congress website.
ISBN: 978-1-5435-9220-7 (library binding)
ISBN: 978-1-4966-6624-6 (paperback)
ISBN: 978-1-5435-9224-5 (eBook PDF)

Summary: Describes what the future may hold in the realm of energy, including technological advancements and human impact.

Image Credits
Alamy: dpa picture alliance, 27; AP Images: Marty Lederhandler, 35 (bottom), Robert F. Bukaty, 21; NASA, 17, 35 (top); Newscom: imageBROKER/Michael Peuckert, 43, ZUMA Press/Connie Zhou, 31; Science Source: CLAUS LUNAU, 39, Peggy Greb/USDA, 25; Shutterstock: Evgeny Vorobyev, 15, Gary Whitton, 7, JoshuaDaniel, 11, Monkey Business Images, 5, NadyGinzburg, 9, Peteri, 29, PopTika, Cover, Ryan Janssens, 23, T.W. van Urk, 13, Wead, 37, yotily, 19, Zbynek Burival, 41; Wikimedia: Olivierabristol, 33

Design Elements
Shutterstock: nanmulti, Zeynur Babayev

Editorial Credits
Editor: Mandy Robbins; Designer: Kay Fraser; Media Researcher: Jo Miller; Production Specialist: Laura Manthe

Printed in the United States of America.
PA100

TABLE OF CONTENTS

Words in bold are in the glossary.

INTRODUCTION
A GROWING NEED FOR POWER

Think of all the things you use that need **electricity**. Electricity turns on the lights. It can be used to heat and cool buildings. It charges cell phones. It runs computers, microwaves, refrigerators, and more.

Yet the use of electricity is fairly new. The first U.S. **power plants** opened in 1882. They each served only a few buildings. In 1925, half of the homes in the United States had electricity. Now few homes are without it.

Electricity changed society in huge ways. It is easier and safer than using candles, oil lamps, or open fires. Many modern devices won't work without it. Our power use grows every year as we get more devices and use them more often. At the same time, the world's population is growing. More people using more technology means a greater need for power. People in the U.S. now use 16 times more electricity than they did in 1950. How can we keep up with this demand?

Look around your kitchen. How many electrical devices do you see?

FACT

Energy cannot be created or destroyed. It can be transferred, and it can change forms. It can also be stored over time.

HOW WE MAKE ELECTRICITY

The sun is a huge source of energy. Plants and animals take in energy from the sun. They store that energy in chemical form. When they die, the energy gets changed to other forms. Some plants and animals that died long ago broke down to become coal, oil, and natural gas. These substances are called **fossil fuels**. They are formed from the remains of living things from long ago. Most of our energy today comes from fossil fuels. That's likely to be true for many years. But we can't count on fossil fuels forever.

Many power plants burn fossil fuels. Doing this converts stored chemical energy to heat energy. The heat is used to boil water and make steam. That's how the heat energy changes into kinetic energy, or the energy of motion. In the form of steam, tiny water droplets race through the air. The moving steam turns a large fan, called a **turbine**. A **generator** then converts the motion energy into electrical energy. Electrical energy can move through wires to wherever we need it.

Power plants create energy people can use, but burning fossil fuels releases smoke that is harmful to the environment.

FACT

Like electricity, natural gas can be sent directly into buildings. Then it may be used for heat or to power appliances such as gas stoves.

CHAPTER 1

WHAT IS JUST AHEAD?

Earth has limited amounts of fossil fuels. People could run out of them in the next 100 years. Plus, burning fossil fuels causes **pollution**. Using fossil fuels is harmful to human health and to the environment. It adds to **climate change**.

Knowing what may happen in the future helps us decide what to do now. Scientists study how the environment is changing. They predict what will happen if we keep using fossil fuels. Then they can suggest changes for a healthier future.

We must switch to different ways of making electricity. We have many alternatives to fossil fuels. Some power plants don't need to burn coal or natural gas. These power plants cause less pollution. They use **renewable energy**, so they will never run out. These forms of energy are growing in popularity.

Most vehicles use gasoline, a product made from oil.
Oil is a fossil fuel.

FACT
Nearly half of Americans live in areas with unhealthy air.
Cars and other vehicles are the worst polluters.

ENERGY INSIDE THE EARTH

The inside of Earth is very hot. In some places, underground water is nearly 1,800 degrees Fahrenheit (1,000 degrees Celsius). **Geothermal** power uses this heat from inside our planet. If the water is close enough to the surface, a well can tap it. The hot water comes up and turns to steam. That steam turns turbines. Making electricity this way skips the step of burning fossil fuels.

Geothermal plants release harmless steam. Most of the water can be pumped back into the ground. There it can pick up more heat and be used again.

Geothermal power is mainly used where hot water comes near the surface. In the near future, deeper wells could reach Earth's heat in more areas. The deepest wells planned now are 2.8 miles (4.5 kilometers) deep. Companies hope to dig even deeper wells soon. Deeper wells mean more heat and more power.

The Ohaaki Power Station in New Zealand taps into local geothermal resources.

Earth's heat sometimes reaches the surface in hot springs or volcanoes. These features show where geothermal resources can be found.

ENERGY FROM THE AIR

Wind energy is another popular option for the future. Wind turbines don't need steam to move them. They use the power of the blowing wind. This saves a step and does not release pollutants from burning fossil fuels. Wind is free and will never run out.

Wind power works best where the wind is strong and steady. The wind is stronger and steadier high above the ground. Taller towers with longer blades can capture this wind. One planned wind turbine will reach more than 850 feet (260 meters) tall. However, tall, skinny things tend to bend in high winds. Blades can twist and destroy the turbine. Engineers must find ways to make the turbines strong enough to withstand strong winds.

Ocean wind is also stronger and steadier than wind on land. It is more expensive to build wind farms in the ocean, but they can produce more power. Many countries are building wind farms near the coast. In the U.S., at least 15 projects are planned. Designers in Massachusetts hope to have 84 offshore wind turbines up and running by 2022.

The Netherlands has wind turbines both on land and offshore.

WORKING WITH WATER

A rushing river also has a lot of motion energy. At a **hydropower** plant, a dam built across a river directs the water into a tunnel. The water turns a turbine to make electricity, and then it flows downstream.

Hydropower supplies electricity to more than 1 billion people around the world. In Canada, nearly 60 percent of electricity comes from water. Norway, Brazil, and the Democratic Republic of the Congo use even more. Ninety percent of their electricity comes from water.

The U.S. has 80,000 dams. Most were built to control flooding or bring water to people and farm fields. Only 3 percent are now used for power. Building new dams is expensive, but hydropower could be added to more of the dams that exist now. Plans are underway to convert 32 U.S. dams to hydropower. Thousands more could be converted in the future. Hydropower makes only about 7 percent of U.S. electricity. That number should grow as hydropower is added to more dams.

The Krasnoyarsk Dam creates electricity from the power of the Yenisei River in Siberia, Russia.

USING THE SUN'S RAYS

Some renewable energies don't use turbines at all. Solar energy uses panels to capture sunlight, a form of heat energy. The panels convert the sun's heat energy to electric energy. Groups of large solar panels can provide power to a town. Small panels can power small devices such as calculators. Solar panels can power the lights on city streets. They can charge electric vehicles in public parking lots. The International Space Station (ISS) even uses solar power.

New companies are experimenting with solar windows and even solar paint. Solar windows capture some of the sun's energy for power but also allow light to pass through. Solar paint could turn almost any surface into a power source. With solar windows and paint, buildings could make power to run lights and machines inside.

FACT
Electric vehicles are usually plugged in to charge. New solar panels can go on cars so they get power from the sun too.

POWERING REMOTE LOCATIONS

In some countries, power plants only provide electricity to big cities. People who live far from cities need another way to get power. Millions of homes in Asia and Africa now use solar panels. Solar panels power schools and health clinics in remote areas.

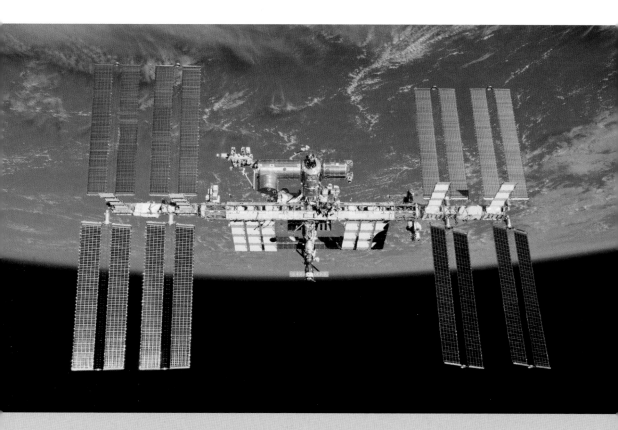

The solar panels on the ISS can collect solar energy for longer periods of time per day than solar panels on Earth can.

ENERGY FROM ATOMS

Nuclear power splits **atoms** to create nuclear reactions. These reactions release energy and make heat. Many countries use nuclear power. In the United States, almost 20 percent of electricity comes from it.

Nuclear power does not pollute the air. However, it does produce deadly waste, which must be stored safely. Many people are afraid of nuclear power. No one wants nuclear waste near where they live, so the waste is piling up. Even worse, a damaged nuclear power plant can release poisons. Large disasters have killed people and polluted big areas.

Nuclear power at its best is clean, efficient, and cheap to produce once the expense of building the plant is covered. Some experts say nuclear power is the best bet for the future.

The latest nuclear plants recycle used fuel. By doing this, they make more energy and leave less dangerous waste. Russia started using the first of these power plants in 2015. Taking steps to solve the waste problem gives nuclear energy a safer future.

The Tihange Nuclear Power Station in Belgium has three nuclear power plants.

CHAPTER 2

WHAT DOES THE FUTURE HOLD?

What happens if we look further into the future? Some forms of renewable energy have potential, but they're not working well yet. The next 30 years may prove whether these ideas take off or fade.

The first offshore wind farm in the U.S. started working in 2016. A dozen more were soon planned. Most projects are on the East Coast.

People in California are interested in offshore wind too. However, the water off the West Coast is deeper. Turbines attached to the ocean floor wouldn't reach above the water to the wind. Floating wind turbines may solve this problem. Currently, 14 companies are designing floating turbines. They would work on the West Coast where the ocean is deep. However, the designs have yet to prove they will work. The earliest tests might start in 2025.

DO LESS HARM

Even renewable energy has downsides, particularly to animals. Wind turbines on land can kill bats and birds. Turbines in the ocean might harm marine animals. We can lessen the damage with careful design and planning. For example, fewer birds die if wind turbines are not built in common travel paths of birds.

The first offshore wind turbine in the United States was installed off the coast of Castine, Maine, in 2013.

RUSHING IN AND OUT

Ocean winds have power, and so do ocean waves. Tidal power uses energy from ocean water. Tidal power can be captured with turbines similar to wind turbines. These turbines are attached to the ocean floor near the coastline, where tides rush in and out. The force of the ocean current turns the blades.

Tidal turbines must withstand ocean storms. They must not be damaged by salty seawater. These are tough challenges. Scotland and South Korea are successfully using tidal turbines. Wave power could produce twice the electricity the world now makes. But it doesn't work well everywhere yet.

In the U.S., the Ocean Power Energy Company built the first commercial tidal power project in Maine. However, the tides were not strong enough to produce the desired power. The company decided to end the project and remove the turbines. Other states on both the West and East Coasts have considered projects for tidal power. None are in progress yet.

People see the power of tidal waves from above. But tidal power turbines would be beneath the water's surface.

FACT

Waves also have energy in the up-and-down motion of the water. Devices that float on top of the water can capture that wave energy.

PLANTS AS FUEL

Biofuel uses matter from living things, such as plants, as fuel. This fuel is often used to power vehicles. It is burned to make steam that turns turbines. It sounds just like using fossil fuels. But this system burns new plants instead of ancient plants. Plants are renewable, so we can quickly grow more of them. Biofuel does release pollution when it is burned. But plants absorb CO_2, the main gas that causes climate change. The growing plants help clean the air.

In the United States, most biofuel is made from corn. In Brazil, many cars run on fuel made from sugarcane. Europe uses a fuel often made from palm oil. But some experts say we should not use food crops for fuel. It uses too much land and water. Turning the plants into fuels takes a lot of energy as well.

That doesn't mean biofuel won't work. College students in Massachusetts found a way to get more energy out of biofuel from food scraps. Restaurants and grocery stores end up with many food scraps and used cooking oil. They could be turned into biofuel.

Engineers Neil Goldberg (left) and Akwasi Boateng (right) built a machine that converts crop leftovers into renewable energy.

POOP FUEL

What else could we use as fuel? How about toilet waste? Wastewater, or sewage, is water left over from other uses, such as baths, laundry, and washing dishes. It also includes what is flushed down the toilet. Wastewater must be cleaned, or it can pollute local water supplies.

Alabama built the world's first algae biofuel system. In this system, plastic bags are filled with wastewater. Algae, which are tiny plants, are added to the bags. The algae feed off the wastewater, which cleans the water. Then the algae are used as biofuel. The cleaned water flows into the local bay. This solves two problems. It cleans dirty water, and it makes cheap fuel.

Some experts thought we would be making billions of gallons of fuel from algae before now. Instead, many algae biofuel companies have failed. While this type of fuel holds so much promise, it is simply too expensive to produce at this point. A few companies are still trying. Synthetic Genomics has partnered with oil company ExxonMobil. They hope to produce 10,000 barrels of algae biofuel a day by 2025.

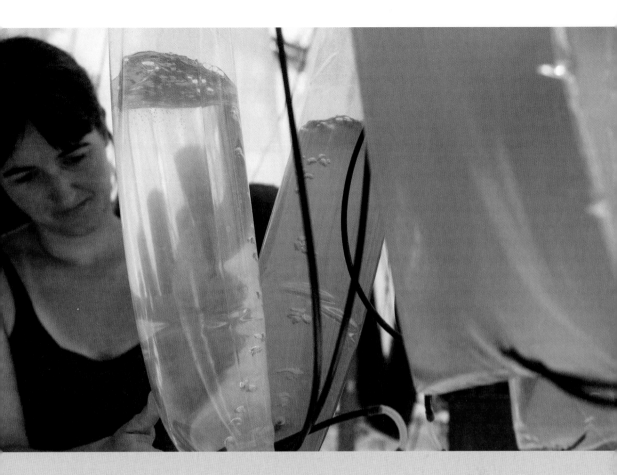

Researchers at the Institute for Bio and Geosciences in Germany are working on making biofuel from algae more efficient.

FACT

Astronauts need to recycle wastewater in space. The United States space agency, NASA is working on a system to clean wastewater with algae.

IS SMALLER BETTER?

People often look for big solutions to big problems. But sometimes the solution is to go smaller. Today most power plants are huge and wasteful. They must send their energy long distances. Doing that is difficult and expensive. In the future, each city and town could have its own small power plant. That way, energy wouldn't be wasted traveling long distances.

With small local power plants, communities could use whatever energy source works best at that location. Even individual homes and businesses could generate part of their own power. Coastal cities and towns could run on tidal power. Towns with geothermal resources could tap that for power. Inland towns in flat areas could run on wind power. Desert locations could use solar power.

The town of Pilsen in the Czech Republic has a solar power station.

USE LESS, SAVE MORE

We can save energy in small ways too. Turn off lights in an empty room. Turn off devices you're not using. Each of us can help in little ways.

Artificial intelligence (AI) can turn small savings into big results. Artificial intelligence uses computers to make decisions. In a building, AI can monitor lights, heat, and appliances. It makes changes to keep people comfortable while using less energy.

Google has many large groups of computers. They need to stay cool or the computers will fail. The company trusted AI to judge how much cooling was needed. Google cut the amount of energy used to cool its computers by 40 percent.

A company called DeepMind made the system Google used. The United Kingdom's power company National Grid is considering using DeepMind to help cut its energy use as well. The whole country might start using AI to control its power. Other countries and companies could follow.

Google data centers, such as this one in Council Bluffs, Iowa, are located around the world.

CHAPTER 3

WHAT IS WAY OUT THERE?

The further ahead we look, the more difficult predictions become. What could the world look like in 50 or 100 years? Scientists and engineers constantly explore new paths. Some ideas seem unlikely. Yet these creative ideas could solve the world's energy problems.

FLYING WIND FARMS

Air currents are stronger and steadier high above ground. But tower turbines have height limits. The company Altaeros Energies has a solution. It is building wind turbines that fly! These turbines could make two or three times the energy of ground wind turbines.

Go high enough, and winds can reach more than 100 miles (160 km) per hour. But that requires reaching a height of 20,000 to 50,000 feet (6,000 to 15,000 m). Reaching those winds brings many challenges. For one, airplanes fly in that range. Flying wind farms would have to stay out of flight paths.

The winds high above the ground could provide 100 times the energy the world needs.

The Kiwee One is a flying, wind-powered generator developed by French company Kitewinder.

SOLAR ENERGY IN SPACE

Solar energy is also more powerful high above ground. In space, it's never cloudy or rainy. Much more solar energy could be captured there. But how do we get the energy to the ground?

Scientists have built one model showing how this could work in the future. A tile would capture sunlight and convert it to electricity. The electricity would then be converted to radio waves. An antenna would send the energy to Earth. A receiver on the ground would capture the waves. They would then be turned back into electricity. Other ideas suggest using microwaves or lasers instead of radio waves.

So far the model has only been tested in a lab. Many questions remain. Will the system be safe? Can the tiles be made light enough? Would it be affordable to get them into space? Currently, it costs thousands of dollars to send 2 pounds (1 kilogram) of material into space. A space-based energy system could cost more than $36 billion to build.

Solar panels only work when the sun is shining. Putting them in space would generate more electricity than they do on Earth.

FACT

Author Isaac Asimov came up with the idea of space-based solar power in 1941. He used it in his science-fiction short story, "Reason."

SUPER VOLCANIC HEAT

Geothermal power already uses heat from underground. Sometimes this heat reaches Earth's surface in a volcano. In the distant future, scientists could figure out how to harness volcanic energy for human use.

A supervolcano is a very large volcano capable of huge eruptions. Our planet has 20 known supervolcanoes. No supervolcano is likely to erupt in the next 100 years. But if one did, it could change the climate. Food crops could fail. People could starve.

One supervolcano is under Yellowstone National Park. NASA wants to stop this volcano from erupting. NASA has proposed turning Yellowstone into a geothermal power plant. Pumping water into the **magma** chamber would cool the volcano. The water could pick up heat and return to the surface. We could use that heat for energy. Once the volcano is cooled enough, it should not erupt.

Italy's Mount Etna erupted in 2014, releasing a huge amount of heat energy. In the future, humans could convert this type of energy to electricity.

CHALLENGES OF VOLCANIC POWER

Using Yellowstone as a volcanic power plant has many risks, though. The most dangerous is that trying to stop the volcano might actually set it off. Drilling into the magma chamber could release harmful gases. It could make the cap over the magma chamber more brittle. Then that cap could crack, releasing lava. Experts at NASA think they can avoid this. They would drill from the lower sides to get under the magma chamber.

But there are more challenges. The magma chamber under Yellowstone is very deep. Workers would need to pump water down 6 miles (10 km). That's much farther than anyone has drilled before.

Cost would pose a challenge too. The project is estimated to cost almost $3.5 billion. Selling the energy would pay for that over time. But this initial money would have to come from somewhere.

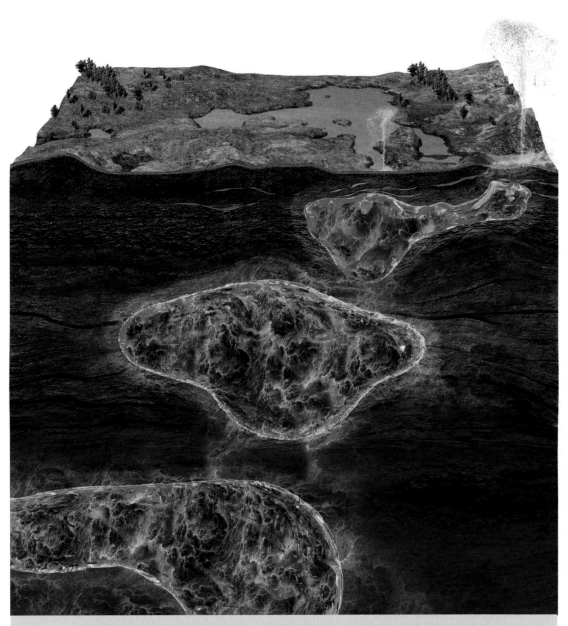

The magma chamber beneath Yellowstone National Park has several different sections.

TINY POWERS

Not every new idea means going into space or deep underground. Small power devices could change the future too.

People are constantly moving. Something called a piezoelectric device makes small amounts of electricity when it is bent or pressed. It can be powered by a person walking. College research labs have made these devices. They are not available for sale yet. Someday we may use such devices to charge our cell phones and smart watches as we move.

Another device makes power through static electricity. Snow carries a positive electrical charge. The snow TENG device is made out of a material with a negative charge. It captures the positive charge from snow. It combines those charges into electricity. Scientists at UCLA made the first snow TENG with a 3-D printer. No company is making the devices to sell yet. But someday snow-powered devices could be added to solar panels. The result would work in sun or snow.

When solar panels are covered in snow, they don't work.
The snow TENG device could solve that problem.

LOOKING AHEAD

Can we stop using fossil fuels? Some places already have. Iceland gets all its energy from other sources. It uses hydropower and geothermal power. Costa Rica and Norway use almost all renewable energy too.

In the United States, Georgetown, Texas, currently gets all its energy from wind turbines and solar panels. The town of 67,000 people is the largest U.S. city using only renewable energy, though the costs have proven challenging. Many other U.S. cities hope to switch entirely to renewable energy. Plans often call for making the switch within 10 to 20 years.

For more than 100 years, many people have said we need to change our energy sources. Inventor Alexander Graham Bell addressed this in a 1917 speech. He thought it might be possible to get energy from the tides, waves, or the sun's rays. Today we can do all that. Only time will tell what the future holds. Scientists continue to make great strides in the area of renewable energy.

Ljósafoss Power Station is one of Iceland's many renewable energy power plants.

TIMELINE

6000 BC People use sails to harness wind energy for transportation.

500 BC The Chinese use natural gas leaking from the ground to boil water.

AD 1100 Windmills in Europe harness wind power to grind grain.

1690 Coal begins to replace wood as fuel in Europe, due to forests being cut down.

1850–1945 The primary fuel source in the U.S. is coal. Natural gas is also used for lighting and wood for heating.

1859 The first oil well in the United States is drilled in Pennsylvania.

1868 The first modern solar power plant is used to heat water for a steam engine in Algeria, Africa.

1882 The world's first commercial hydropower plant starts in Wisconsin.

1888 The first windmill used to generate electricity is developed in Ohio.

1925 Half of homes in the United States have electric power.

1935 Hoover Dam, the world's largest hydropower plant, is built.

1938 German scientist Otto Hahn discovers the process for nuclear energy.

1950 Oil-based fuels become the most used fuels in the United States.

1951 The first nuclear power plant is built in Idaho.

1974 The solar-cell device is developed in the U.S.

1980 The world's first wind farm is built in New Hampshire.

1981 The first large-scale solar power plant starts working in California.

2014 Alabama builds the world's first algae biofuel system.

2016 The first offshore wind farm in the U.S. starts working.

GLOSSARY

artificial intelligence (ar-ti-FISH-uhl in-TEL-uh-junss)—the ability of a machine to imitate human behavior

atom (AT-uhm)—an element in its smallest form

biofuel (BYE-oh-fyoo-uhl)—a fuel made of, or produced from, plant material

climate change (KLY-muht CHAYNJ)—a significant change in Earth's climate over a period of time

electricity (i-lek-TRISS-uh-tee)—a natural force used to make light and heat or to make machines work

fossil fuel (FAH-suhl FYOOL)—natural fuel formed from the remains of plants and animals

generator (JEN-uh-ray-tur)—a machine used to convert mechanical energy into electricity

geothermal (jee-oh-THUR-muhl)—heat inside the earth

hydropower (HY-druh-pow-uhr)—the production of electricity from moving water

magma (MAG-muh)—melted rock beneath Earth's crust

nuclear (NOO-klee-ur)—describes energy converted by splitting two particles of matter

pollution (puh-LOO-shuhn)—harmful materials that damage the air, water, and soil

power plant (POW-ur PLANT)—a building or group of buildings used to create electricity

renewable energy (ri-NOO-uh-buhl EN-er-jee)—power from sources that will not be used up

turbine (TUR-bine)—a machine with blades that can be turned by a moving fluid such as steam or water

READ MORE

Green, Dan. *Energy.* New York: DK Children, 2016.

Hawbaker, Emily. *Energy Lab for Kids: 40 Exciting Experiments to Explore, Create, Harness, and Unleash Energy.* Beverly, MA: Quarry Books, 2017.

Sneideman, Joshua. *Renewable Energy: Discover the Fuel of the Future With 20 Projects.* Norwich, VT: Nomad Press, 2016.

INTERNET SITES

Alliant Energy Kids
www.alliantenergykids.com/

Science Kids: Energy Facts
www.sciencekids.co.nz/sciencefacts/energy.html

The U.S. Energy Information Administration: Energy Explained
www.eia.gov/energyexplained/

INDEX

CONTENTS

ACKNOWLEDGMENTS

There are many sterling leaders who have made this book possible.

Ruthmary Wood of the Aurora Public Library was instrumental in finding historical facts for me, while Dave Chrestenson, Greg Stangl, Donnell Collins, Ron Stewart, and Photography by Feltes provided refreshingly beautiful photographs. The other photographs used in the book were provided by each of the profile subjects unless otherwise indicated. The Aurora Historical Society and the Community Foundation of the Fox River Valley provided several images as noted.

Grateful thanks are given to Mayor Tom Weisner and the Aurora City Council, Ruth Dieterich Wagner, Dr. Christine J. Sobek, Sharon Stredde and the Community Foundation of the Fox River Valley, my Arcadia editors Tiffany Frary, Erin Vosgien, and Laura Saylor, Daniel D. Dolan Sr., John Jaros, Ruthmary Wood, Diane Christian, Jeff Scull, Jeff Noblitt, Adam Punter, Mayor David Pierce, Mayor Albert McCoy, Christina Campos, Justine Kopykov, Carter Crane, Jason Crane, Dr. Bill Marzano, Diane Piccioulo, Faith and Lloyd Jones, Al Benson, Mary Clark and Neal Ormond, Marissa Amoni, Dave Chrestenson Sr., Dorothy Burns, Marissa Happ, Dana Peterson, Rev. Cyndi Gavin, Wayne Johnson, Mary Jane Hollis, Ann Humiston, Dee Nila Basile, Nancy McCaul, Nancy and Jim Hopp, Wendell Minor, Kathleen Fennell, Mary Covelli, Lorna Ruddy, John Ruddy, Philip and Colleen Ruddy, Georgette Prisco, Martha Prisco, Robert Prisco, Joyce Dlugopolski, Sheila Scott-Wilkinson, Karen Shaffer, Jerry Lubshina, John Schaefer, Mary Ann Schaefer, Mike McCoy, Mavis Bates, Gary and Mary Roberts, Fr. David Engbarth, Art Stiegleiter, Kay and Steve Hatcher, Beth Zanis, MaryAnn Andrews, Cecelia Sanders, Marlis Robillard Marrello, Kenlyn Nash-Demeter, Bob Bonifas, Mike Leonardi, Jim Leonardi, Susan Palmer, Jim Oberweis, Jo Ann Suhler Bailey, Gladys Larson Mason, Barbara Dray, Jane Regnier, Maureen Granger, Ryan Reuland, Jeff Reuland, Brigit Reuland-Foster, Jacqueline Ahasic, Doug Lima, Bill Grabarek, Nancy Huntoon Garbe, Carleton Huntoon, Bob Brent, Delbert Peterson, Joseph Henning, Lillian Perry, Lori Renzetti, Wynette Edwards, Bob Mangers, Jan Mangers, Arlene Shoemaker, Theresa Shoemaker, Sarah and Senator Chris Lauzen, Perry Slade, Willie Etta Bonner Wright, Geraldine Pilmer, Mary Lou Chapa, and Mary Ann Kutnick.

Thank you also to the following for providing information through their archives and resources: Aurora Historical Society; Aurora Public Library; Aurora Regional Fire Museum; Aurora University; Community Foundation of the Fox River Valley; Montgomery Public Library, and Waubonsee Community College, the *Chicago Daily Tribune*, *The Voice*, the *Chicago Tribune*, and the *Beacon-News*.

LEGENDARY LOCALS
OF

AURORA

ILLINOIS

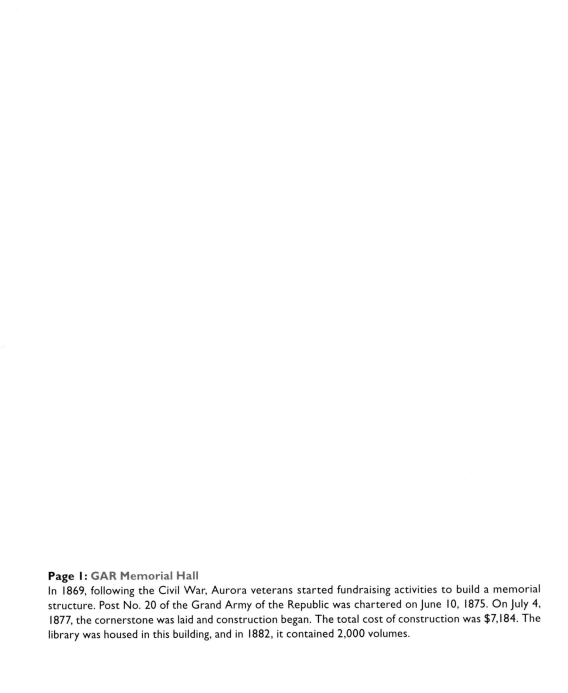

Page 1: GAR Memorial Hall
In 1869, following the Civil War, Aurora veterans started fundraising activities to build a memorial structure. Post No. 20 of the Grand Army of the Republic was chartered on June 10, 1875. On July 4, 1877, the cornerstone was laid and construction began. The total cost of construction was $7,184. The library was housed in this building, and in 1882, it contained 2,000 volumes.

LEGENDARY LOCALS
— OF —

AURORA
ILLINOIS

JO FREDELL HIGGINS

LEGENDARY
LOCALS

Copyright © 2012 by Jo Fredell Higgins
ISBN978-1-4671-0035-9

Legendary Locals is an imprint of Arcadia Publishing
Charleston, South Carolina

Printed in the United States of America

Library of Congress Control Number: 2012938519

For all general information, please contact Arcadia Publishing:
Telephone 843-853-2070
Fax 843-853-0044
E-mail sales@arcadiapublishing.com
For customer service and orders:
Toll-Free 1-888-313-2665

Visit us on the Internet at www.arcadiapublishing.com

Dedication
To my fraternal grandmother, Mayme Smith Fredell.

On the Cover: From left to right:
(Top row) Daniel D. Dolan Sr., Dolan & Murphy real estate businessman (Courtesy of Donnell Collins, page 97); Dr. Christine J. Sobek, president of Waubonsee Community College (Courtesy of Waubonsee Community College archives, page 87); Katherine Mateas and baby Harry, c. 1917, residents (Courtesy of Mateas family, cover image); Mayor Thomas J. Weisner, City of Aurora (Courtesy of the mayor's office, Aurora, page 71); Ruth Dieterich Wagner, age 100 and beloved community leader (EAHS photograph, page 117).
(Middle Row) Maud Powell, noted violinist (Courtesy of the Maud Powell Society for Music and Education, Karen A. Schaffer, page 35); Joseph Stolp, city founder (Courtesy of the Aurora Historical Society, page 10); Marie Wilkinson, celebrated civic leader (Courtesy of Sheila Scott-Wilkinson, page 52); Mary Jane Hollis, accomplished city leader (Courtesy of Mary Jane Hollis, page 65); Wendell Minor, published artist and award-winning illustrator (Courtesy of Wendell Minor, page 33).
(Bottom Row) Ruth Van Sickle Ford, renowned artist (Courtesy of Nancy Smith Hopp, page 38); Dr. Bernard J. Cigrand, father of Flag Day (Courtesy of the Aurora Historical Society, page 12); Nancy Smith Hopp, author and civic leader (Courtesy of Nancy Smith Hopp, page 34); Bruno Bartoszek, World War II military leader and community volunteer (Waubonsee Community College archives, page 51); Grace Nicholson, respected educator and school principal (Courtesy of Veronica Radowicz, page 92).

FOREWORD

Aurora, the "City of Lights," welcomed its first commercial enterprise, the McCarty sawmill, on June 12, 1835. The Aurora Woolen Mills, owned and operated by J.D. Stolp, was one of the most important establishments during that period. A writer in 1910 suggested that Aurora "has a quiet resident people, an order-loving population, a cultured and refined people." The population had grown to 35,000 by 1910, with 35 miles of paved streets.

Aurora now has a population of nearly 200,000, which is diverse, energetic, and multicultural. The Hispanic population is 42 percent, with immigrants also from China, Peru, Greece, Spain, Germany, and 25 other countries comprising the neighborhoods and strengthening the churches.

Throughout the history of the City of Aurora, there have been many leaders who have contributed to the fabric and the quality of daily life. As Aurora celebrates her 175th year, we are reminded of these august individuals. This book will offer a composite portrait of those, past and present, who have made the city of Aurora "second to none."

This portrait is written by noted writer Jo Fredell Higgins. It is a remarkable photographic history for future generations as they, also, discover Aurora.

—Tom Weisner, mayor

INTRODUCTION

Over the 175 years of its existence, Aurora, our bright City of Lights, has been shaped by thousands of individuals. Author Jo Fredell Higgins has taken on the task of presenting leaders who have made a significant impact. The author has accomplished this through in-depth research as well as first-person interviews, highlighting individuals from the pioneers of yesterday to those who are influencing our great city today. These include hardy settlers, humble shopkeepers, and average factory workers, as well as artists, poets, and teachers who have helped shape this community.

Native Americans occupied this area when the McCarty brothers arrived in 1834 to build mills on the Fox River, around which Aurora would grow. The railroad came in the 1850s and helped Aurora grow as a center of manufacturing and commerce. By 1910, the population had reached nearly 30,000, with 26 schools and 42 churches. The 2010 census showed Aurora with nearly 200,000 residents, making it the second-largest city in the state.

During those years, Aurora's luminaries have included pioneer Joseph Stolp, whose woolen mills were Aurora's first major industry; Maud Powell, the premier female violinist of the late-19th and early-20th century; and controversial 1950s mayor Paul Egan, called by *Chicago* magazine "The Worst Mayor in America."

Over the decades, successive waves of immigrants have left their mark on Aurora. Today, Hispanics comprise more than 40 percent of the present population and are well represented here. Hector Jordan was the first Hispanic police officer in Aurora. State senator Linda Chapa LaVia has been the first Hispanic Democrat outside of Cook County to be elected to the state legislature. Lourdes Blacksmith is director of government and multicultural affairs at Waubonsee Community College. Christina Campos is the Aurora Township supervisor. Sheriff Pat Perez has been a prominent public servant, with family ties to Aurora's earliest Hispanic residents.

In *Legendary Locals of Aurora*, you will read about those who have contributed to their fellow man in every aspect of Aurora society. Their inspiring narratives will illustrate why they were chosen for this honor. Enjoy a journey through time with the illustrious citizenry of yesterday and today.

—John Jaros, executive director
Aurora Historical Society

CHAPTER ONE

Settlers and Immigrants

Imagine standing inside a cathedral when every ray of light reveals a harmony of unspeakable splendor.

—*Nathaniel Hawthorne*

Samuel and Joseph McCarty
The founding fathers of Aurora are Samuel and Joseph McCarty. In April 1834, Joseph laid claim to 860 acres on both sides of the Fox River. In November 1834, his brother Samuel joined him and they began plotting out streets and land parcels. Samuel met Chief Waubonsie, whose wigwams were set close by the Fox River. Samuel told him he was planning to build a mill to grind their *quashgen*, or meal. Later, the government moved the tribe to Council Bluffs, west of the Mississippi River, with the gift of a 40-mile-square parcel of land. The McCartys held the first meeting of the Methodist church in their home in the autumn of 1837. Whale-oil lamps lit the rooms. In the late 1840s, Samuel constructed the first water system along the foot of a bluff where several springs flowed out of the ground. On December 27, 1874, the First Methodist Episcopal Church was dedicated; construction cost $60,000. Joseph had moved to Alabama for his health, and he died there in the spring of 1839 at about 31 years old. Samuel McCarty is shown in this photograph from the Aurora Historical Society archives. There is no known photograph of Joseph McCarty.

Joseph and Frederick Stolp
Arriving from Pulneyville, New York, on June 12, 1837, Joseph Stolp built his carding mill on the banks of the Fox River. He was born on August 16, 1812. He wed Temperence Dustin of Naperville, and they had five children. Uncle Frederick Stolp had arrived in 1834. Frederick sold the land to his nephew Joseph, who paid $12.72 for the 10.18 acres—a rate of $1.25 an acre—for Stolp's Island on January 8, 1842. Joseph complained that the government paid the Indians 4¢ an acre for the land, but he had to pay $1.25! Joseph donated the land for the GAR Memorial building and city hall on Fox Street, now called Downer Place. Their sixth-generation descendant living in Aurora is Jo Ann Suhler Bailey, who graduated from West Aurora High School in 1966. She has one son, James. She retired from Farmer's Insurance in 2006 after 36 years. Her great-grandfather had three daughters—Mabel, Lena, and Clementine—and no sons, so the Stolp surname ended with that generation in 1919. (Courtesy of the Aurora Historical Society archives.)

Postmaster Burr Winton
Aurora's first postmaster
was Burr Winton, a
Democrat, appointed by
Martin Van Buren. He held
the office for 10 years. The
date of the city's birth is
March 2, 1837, because
McCarty Mills was then
renamed Aurora, and the
federal post office was
opened. In 1838, postmaster
Winton helped form an
association, with members
paying $2 a share, to found
a town library. Eventually,
$100 was raised, and Burr
became the first librarian,
keeping the books in his
home. The group was named
the Young Men's Association,
and it accumulated 600
volumes. After the Civil
War began, the association
disbanded. The Aurora
Library Association took
over in 1865, and the book
collection was kept in the
rear of the post office in
city hall, rent-free. Winton
is buried at the Aurora
Root Street Cemetery
with the epitaph, "Gone
but remembered with the
fondest affection." His death
date is December 14, 1885.
(Courtesy of the Aurora
Historical Society archives.)

Burr Winton,
First Post Master.

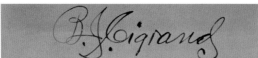

Dr. Bernard J. Cigrand
The founding father of Flag Day, Bernard Cigrand was the son of Nicholas and Susan Cigrand, immigrants from Luxembourg. He was born in Wisconsin on October 1, 1866. In 1912, the Bernard Cigrand family moved to Batavia, and in 1920, he opened a dental office at 47 1/2 Fox Street, in Aurora. During World War I, Cigrand served with distinction as a lieutenant in the Navy. After years of letter-writing from his rolltop desk, Dr. Cigrand saw a national day to honor the American flag become official. Pres. Woodrow Wilson in 1916 had issued a proclamation calling for a nationwide observation of Flag Day on June 14. In 1949, Pres. Harry Truman made it an official, permanent day of observance. Dr. Cigrand died of a heart attack on May 16, 1932. Blackberry Farm–Pioneer Village in Aurora is now home to an exhibit that reproduces the dental office of Dr. Cigrand. (Courtesy of the Community Foundation of the Fox River Valley archives.)

Annunzio Covelli

Born in the foothills of Cosenza, Italy, on January 1, 1927, Annunzio (the tallest boy in the photograph with sisters Franchesca and Ida and brother Armando) was one of four children by Filiberto and Emilia Covelli. He came to the United States in 1953 and settled in Aurora. He was sponsored by his Aunt Ann and Uncle Mike Covelli, who owned and operated the Rivoili restaurant on River Street. He met Florence DeBartolo, who was an English teacher with Aurora's West District No. 129. They wed on August 11, 1954, and had two children, Marianne and Michael. On September 30, 1955, Covelli purchased the Sinclair gas station on North Avenue and Broadway, and he operated a newer location at Randall Road and Galena Boulevard for over 20 years. His son, Michael, joined the business in 1977, and his son-in-law, Gordon Murray, did so in 1978. In 1988, Michael left the business, and Covelli turned over the operation to Gordon, who bought it shortly after that. Covelli's death date is December 4, 1999. Florence died on June 15, 2010.

DeBartolo Family

Michael DeBartolo was born on June 6, 1885, in the province of Cosenza, Italy. He was an only child. He wed Louisa Molinaro in Italy, and they immigrated to the United States in 1912 aboard the *Princess Irene* from Naples, Italy. The family moved to Aurora in 1925. He started a grocery business at 519 South Lake Street. He served on the Aurora Zoning Commission and was a trustee of Holy Angels Church. He died on February 19, 1945. Louisa Molinaro was born on May 2, 1884, in Cosenza to Ignatio Molinaro and Anunziato Rinneli. Louisa made altar linens for Holy Angels Church and raised their 10 children. She died on February 9, 1960. Shown in this early-1920s family photograph of the DeBartolos are, from left to right, (first row) Florence, Louisa, Elsie, Michael, and Yolanda; (second row) Oliver, Hansel, Ida, Chuck, Elvera. Two children not shown in the photograph are Helen and Ann. An 11th child had died in Italy. (Courtesy of Marilyn Spieler.)

Casper Ahasic Sr.
Ahasic was born in Slovenia in 1890 and immigrated to America when he was 17 years old. His first job was at the rock quarry in Lemont. He then walked from Lemont to Aurora, where he found work at the Burlington Railroad. In 1919, he began a wet wash in the basement of his home on Pierce Street, calling it Aurora Laundry Co. The laundry would deliver wet clothes back to homes so women could hang them out on a clothesline and then do their ironing. In 1975, another branch called the Valley Linen Supply began at 562 South River Street. He died in 1987 at age 97. Today, the company has third- and fourth-generation family members handling service mainly for the hospitality industry and health-care facilities. Thus, 93 years later, its 75 employees work one shift daily, with Leo, Barb, Mike, and Jacqueline Ahasic continuing the fine tradition established by first-generation owner Casper Ahasic Sr. Shown is Casper and his wife, Antoinette. (Courtesy of Jacqueline Ahasic.)

Hector Jordan
Special Agent Jordan was born in El Paso, Texas, and came to Aurora at the age of 16. Jordan attended East Aurora High School. On April 16, 1956, he was sworn in as the first Hispanic Aurora police officer and would serve until November 22, 1965. He received Aurora's first Policeman of the Year Award in 1964. He then took a position as a special agent with the Federal Bureau of Narcotics, today's Drug Enforcement Administration. On the eve of taking a federal appointment in Spain, Jordan was critically beaten by seven men when he came to the aid of a fellow police officer. The date was September 20, 1970. Jordan was 38 years old. The city of Aurora honored his legacy by naming College Avenue "Hector Jordan Way." Robert Kennedy had written of another that, "He sends forth a tiny ripple of hope and crossing each other form a million different centers of energy and daring." (Courtesy of Dee Basile.)

Michael J. Nila
Senobio and Augustine Nila came to Aurora from Mexico in the 1920s as part of the boxcar community. The family eventually found homes on Indian Avenue near Farnsworth. The street has been given the honorary name Nila Avenue. Michael Joseph Nila (b. 1952) is a retired Area 2 commander on the Aurora Police Department. He is now a senior consultant for Franklin Covey and Guardian Quest, working with police agencies across the country to design and deliver training and consulting solutions. Nila is past president of the Exchange Club of Aurora and has received numerous awards and recognitions throughout his career. He attended St. Nicholas grade school, Marmion Military Academy, and graduated from Aurora Central Catholic in 1970. He holds a bachelor of arts degree in criminal justice management as well as a master of business administration degree from Aurora University. His uncle is Hector Jordan, and he was the grandson of Augustine Nila. (Courtesy of Dee Basile.)

Lourdes "LuLu" Blacksmith

LuLu Romero Blacksmith grew up in Morelia, Mexico, in the Michoacan province. At age 16, she immigrated to America and found work on the third shift in a West Chicago factory. She began taking English as a Second Lesson (ESL) classes and met her husband, Steve Blacksmith, at Ball Horticultural. The family moved to Wisconsin and had daughter Nikki, then returned to the Fox Valley and had daughter Stefanie. Blacksmith then earned her associate degree with honors in 1997 from Waubonsee Community College. She also holds a bachelor's degree in multicultural relations from DePaul University. She became the Hispanic liaison at Provena Hospital. In 1999, she founded Companeros en Salud (Partners in Health), which assists Hispanic citizens to access health care and social services. In 2002, she was asked to join Rep. J. Dennis Hastert's staff. From 2002 to 2008, she was community relations director and district press secretary for Rep. Dennis Hastert, speaker of the US House of Representatives. In 2008, Blacksmith became the director of governmental and multicultural affairs at Waubonsee. She was chair of the board of Fox Valley United Way, a board member of Mutual Ground, and serves on the Dominican Literacy Center Advisory Board. She is currently working toward her doctoral degree in leadership in higher education and organization change at Benedictine University in Lisle.

Lillian Perry (RIGHT)

For her grassroots activism and advocacy for universal civil rights, Lillian Perry (b. 1935) was honored as one of Waubonee Community College's Fabulous 40 alumni in 2006–2007. A native of Georgia, Perry moved to Youngstown, Ohio. After graduation from high school, she came to Illinois to be a governess for wealthy families around Chicago. Mayor Jack Hill asked Perry to be his special assistant for community outreach. She was pivotal in helping create the Kane County Health Department and other community action organizations, including the Citizens for Neighborhood Improvement. Perry worked at Sealmaster Bearing in Aurora for 30 years and now works for state representative Linda Chapa LaVia as director of education, senior citizen, and cultural liaison. Aurora mayor Thomas Weisner declared March 9, 2007, Lillian Perry Day, and the Illinois House of Representatives also recognized her achievements. Her daughter Darlene Rosanna North was born in 1958 and passed away in 2011. (Author collection.)

Gonzalo Arroyo
Arroyo was born in Michoacan, Mexico, on March 5, 1955, and immigrated to the United States in 1980. He holds a master's degree in liberal studies with an emphasis in leadership from North Central College. Arroyo has served on the East Aurora School Board and as director of the Mexican Fest and president of the Bilingual Parent Association. He has been a board member for Two Rivers Head Start, for the Illinois Coalition for Immigrant and Refugee Rights, Emmanuel House, and New Futuro. Arroyo provides technical assistance and participates in international events concerned with Mexico and US relations. He is the cofounder and first president of the Federacion de Clubes Michoacanos en Illinois, and has served on the steering Committee for the Illinois Latino Policy Forum and the National Alliance for Latin American and Caribbean Countries. Arroyo was president of the Aurora Soccer Official Association and American Soccer League Officials, from 1983 to 1996. Since 1996, he has been the director of Family Focus Aurora, a not-for-profit family support organization that focuses on nurturing children and strengthening families. His children are Mariana, Paloma, and Tonatiuh.

Dr. John Struck

John Struck was born in Waukesha, Wisconsin, on September 4, 1945, to John and Vivian Struck. He received his doctor of education degree from Northern Illinois University (NIU) in DeKalb in 1994. He was associated with East Aurora District No. 131 from 1972 to 2005 and was associate superintendent for educational administration from 2002 to 2005. He currently is an assistant professor in the Educational Leadership Program at Aurora University. Dr. Struck was named Educator of the Year for Kane County in 2003, and was nominated for the Marcus and Mark H. Trumbo Excellence in Teaching award in 2009. From 1968 to 1970, he served in Vietnam and received the Combat Infantry Badge, Army Commendation Medals for Valor, Bronze Star Medals for Valor, three Purple Hearts, and the Vietnamese Cross of Gallantry. He has served on the board of Family Focus in Aurora since 1990 and was on the Aurora Township Youth Commission from 1984 to 2000. He is the father of Jennifer, Heather, and John, who passed away at the age of 20. He has four grandchildren.

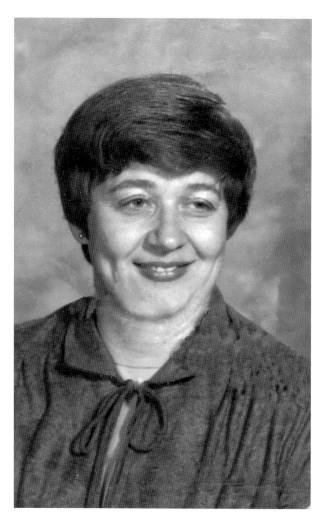

Dr. Anna Sanford

Anna Sanford earned her bachelor of arts degree from George Peabody College in Nashville, Tennessee, and her master of science degree from NIU in DeKalb. She received her doctorate from Purdue University in 1977. Dr. Sanford began teaching in the Aurora East schools in 1966, then was a school principal, assistant superintendent for curriculum and instruction, and administrative assistant for research and development, retiring in 2001. She was president of Delta Kappa Gamma from 1990 to 1992, and served as foundation chair of the Illinois Reading Council Foundation for Literacy and as state coordinator of the International Reading Association. She received the Kane County Administrator of the Year award in 1982 and the Charlotte Danstrom Woman of Achievement award in 1989. She currently serves on the board of Family Focus Aurora and volunteers with the Child Welfare Society and the New England Congregational Church.

Cecelia Sanders (RIGHT)

Born in Michigan in 1943, Sanders moved to Illinois after graduation from Calumet High School. She graduated from Waubonsee Community College and then earned a bachelor of liberal studies from the University of Iowa. She has worked with Senior Services Associates and the Salvation Army Golden Diners, and she is currently the marketing coordinator for Sunnymere of Aurora. Sanders wrote the award application for the 2001 Governor's Award for Unique Achievement for the Salvation Army Golden Diners program. She is married to James K. Sanders, a veterinarian, and they have five children and five grandchildren.

Pastor Reverend Steve C. Zanis (RIGHT)

Steve Zanis was born on December 4, 1921, in Freeport, Illinois, to Christos and Matrona Zanis, who were from the island of Samos, Greece. He graduated from Freeport High School in 1940. After graduation from Pomfret/Holy Cross Seminary (1940–1946), he married Maria Kavallieros in 1947, and they had three children, Alexis, Ellen, and Lucas. Reverend Zanis was ordained in 1947 and served in Seattle, Nashville, and Chicago before becoming the founding priest at St. Athanasios Greek Orthodox Church in 1965. He died on May 10, 1993.

George Nicholas Andrews

A true American success story is that of George N. Andrews. He was born to Nikolaos and Elene Andrutsakis in Gargalianus, Greece, on March 29, 1911. The family came to America in 1918 and all became citizens. Andrews married Demetra "Jennie" Pantazoplos on August 18, 1946, and they raised four children, Elene, MaryAnn, Peter, and Dean. Andrews was a sergeant with the US Army/Air Force, 101st Airborne Division, earning a Purple Heart during World War II. After the war, he took over Main Hatters, which his father had opened in 1919 in downtown Aurora. He helped found the Aurora Boys Baseball league and was inducted into the first hall of fame for Aurora Boys Baseball. One of the baseball fields at Garfield Park was named for him. On June 9, 1974, the city proclaimed it George N. Andrews Day. He was president of the Oak Park School PTA, served on the East Aurora High School Advisory Board, and was a member of the Optimist Club. He and Jennie were charter members of the St. Athanasios Greek Orthodox Church, and he served as its first president. Andrews had a kind heart and helped many Greek immigrants coming to the area find employment, gain citizenship, and assimilate into society. He passed at Rush Presbyterian Hospital in Chicago on December 8, 1981.

Lorraine and Herm Martens
Jacob and Mary Menhardt were born in Yugoslavia. When they landed at Ellis Island, they did not speak English and were almost penniless. They were wed on July 4, 1915, as shown in this photograph, and both became citizens and learned to read English. They had two daughters, Julie and Lorraine (b. 1924). Lorraine moved to Aurora in 1980. She and Herm Martens (b. 1921) met on a blind date and were married on July 15, 1950. They had three sons, Tom, Don, and Bruce—all graduates of the University of Illinois. Herm Martens worked at Meyercord in Chicago as a lithograph artist, retiring in 1985. In 1993, he was named Kiwanian of the Year by the Aurora Golden K. Lorraine became an Adult Literacy Project volunteer and received the Secretary of State Spotlight on Service Volunteer Award for outstanding service in 2006 for tutoring in the GED classroom at Waubonsee Community College. Mayor David Stover recognized Lorraine in 2000 with the Mayor's Certificate of Special Recognition for her "generous and dedicated volunteer" service to adult literacy students. The couple has nine grandchildren.

Robert Bonifas Family

Theodore Bonifas emigrated from Nospelt, Luxembourg, coming to Aurora in 1901. He had been born on September 26, 1876. His profession was blacksmithing, and he worked with the Western Wheeled Scraper Works, later named Austin-Western. He married Mary Walters in April 1903, and they had six children. This photograph shows their youngest son, Arthur (second from left), and his bride, Evelyn May (center), on May 25, 1935. Arthur worked for Oberweis Dairy before buying a neighborhood grocery store located at 508 Grove Street in 1942. In 1968, Arthur and his family built a new supermarket at 1100 Church Road named Art's BI-LO Supermarket. Later in 1968, with competition making the Bonifas supermarket unprofitable, Arthur's oldest son, Robert, and his wife, Alice Hettinger Bonifas, decided to enter the burglar- and fire-alarm business. Alarm Detection Systems is now the 22nd-largest alarm company in the United States, among some 14,000 total alarm companies. It employs 220 and is headquartered at 1111 Church Road in Aurora. Their five children are Steven, Edward, Dale, Connie, and Kimberly. As Ralph Waldo Emerson wrote, "I wish the days to be as centuries, loaded, fragrant."

Louis and Beverly M. Leonardi

Louis (1913–1996) was born on March 11 in Pariana, Province of Lucca, Italy. At the age of 11, his family moved to Chicago and then to St. Charles, where they owned a confectionery called the Sweet Spot. It was there that he met Beverly Pope, and they were wed on October 14, 1933. Louis served with the US Navy, returning home in 1945. He opened Leonardi Refrigeration Service in 1947. He was a member of St. Peter's Church, the American Legion, Amvets, Knights of Columbus, and the Italian-American Club. The couple had three children, Lou Jr., James, and Janis, and 13 grandchildren. Beverly (1915–2012) attended Geneva High School and served as secretary-treasurer of Leonardi TV, Furniture and Appliance. She was a member of the Fox Valley Organ Keyboard Club and was a lifelong bowler. She was inducted into the Aurora WBA Hall of Fame in 2003. Leonardi's new location will be at the corner of Prairie and Lake Streets in Aurora, led by several family members, including Jim and Dan Leonardi.

Melinda Lloyd James
James was born in Campbellsville, Kentucky, in 1960 to Frances Thompson and Billy O'Neil Lloyd. In 1981, she graduated from Murray State University and moved to the Chicago area, receiving her master of science degree in counseling psychology from George Williams College. James was hired by Waubonsee Community College as a counselor in 1989. In 1995, she became an administrator and now serves in the role of assistant vice president of student development. James has two children, Alison and Ben.

Marlis Robillard Marrello
This photograph shows Emma Hartenstein in 1908. Her granddaughter Marlis was born on May 1, 1959, and the family moved to Aurora in 1968. Marlis Robillard graduated from West Aurora High School in 1977 and earned an associate degree in early childhood education from Waubonsee Community College in 1983. Marlis married Richard Marrello on August 30, 1997, and they are raising three children, Mike, Kristie and Derrick. They have seven grandchildren. Richard is a Marine Commandant of the Fox Valley Marine Corps League. They both participate in Operation Welcome You Home, the Patriot Guard Riders, and the Warrior Watch Riders.

Dr. Susan Lee Palmer
Born in Hartford, Connecticut, on May 15, 1949, to Donald and Edith Lay Palmer, Dr. Palmer graduated from Manchester High School in 1967 and wed Richard Westphal in Aurora on May 27, 1978. Palmer received her doctorate in history from NIU in 1986. She was a professor of history, and Richard was a professor of English at Aurora University. She currently serves as curator of the Jenks Memorial Collection of Adventual Materials and as the curator of the Doris K. Colby archives. She has been a board member, vice president, and president of the Aurora Historical Society. Dr. Palmer has won the Gold Ivy Leaf Award for Academic Excellence at Aurora College in 1971 and the Academic Advisor of the Year Award in 1984. (Courtesy of AU archives.)

Schingoethe Center at Aurora University
Herbert and Martha Schingoethe commissioned the building of Dunham Hall at Aurora University, which opened to the public in 1990. They donated their collection of more than 6,000 items of Native American arts, artifacts, and related materials, and they provided major support for the center. Martha Dunham Schingoethe died on February 17, 2004, and Herb followed on March 18, 2005. Herbert and Martha were inducted posthumously into the Fox Valley Arts Hall of Fame, class of 2010. (Courtesy of AU archives.)

Jane Koenig Regnier

Born in Kingston, New York, on August 3, 1948, Jane Regnier and her twin brother, Paul, were the youngest of six children. Jane's mother, Anne, a nurse, was widowed when Jane was three years old. Jane graduated in 1970 from Marquette University, where she met Peter Regnier, a native Auroran. They wed on October 3, 1970, in Rhinebeck, New York. Peter is a Vietnam War veteran and holds a master's degree in business. In 1975, they settled in Aurora to raise their four children, Patrick, Anne, Jim, and Mary Ellen. Jane Regnier took Waubonsee's interpreter training program and obtained professional certification, and she was hired by the college as a sign language interpreter and eventually became the interpreter manager. She was instrumental in expanding the deaf education program to begin serving the needs of all disabled students. She then developed and managed Waubonsee's new distance education program. As technology advanced, Regnier stayed on its cutting edge, obtaining a master's degree in education via online classes at the very beginning of this method of education delivery. Regnier became manager of distance education and was promoted to assistant vice president of program development and distance learning. She was a founding board member of Mutual Ground, a shelter for women and children, led for decades by executive director Linda Healy. She has served on the boards of SciTech Children's Museum and the Aurora Public Art Commission. The Regniers have five grandchildren.

Nicholas and Mary Huberty Born in 1894 in Luxembourg, Nicholas Huberty arrived in Aurora at age 19. He married Mary Margaret Lorenz on May 3, 1915. Their four children were Lawrence, Ruth, Alice, and Richard. Alice had her First Communion in May 1926 at St. Nicholas Church in Aurora. Alice wed Ralph Baumann on May 10, 1947, at St. Nicholas Church. Ralph joined the Coast Guard at age 18 and retired in 1968 after 20 years of service. They had two children, Patricia and Gary, three grandchildren, and four great-grandchildren. Alice completed her GED and received her diploma at age 58. They are members of St. Joseph's Parish. Ralph worked as a mechanic and service repairman before retiring in 1967. (Courtesy of Wayne Johnson.)

Dr. Stephanie Pace Marshall
Marshall was born on July 19, 1945, to Domnick and Anne Pace Marshall in the Bronx, New York, and graduated from East Meadow High School on Long Island in 1963. She received her bachelor of arts degree at Queens College in Flushing, New York, received her master of arts degree from the University of Chicago, and earned her doctorate from Loyola University in Chicago. Dr. Marshall has been awarded honorary degrees from Dominican University, Aurora University, North Central College, and Illinois Wesleyan University. She has been the president emerita of the Illinois Mathematics and Science Academy (IMSA) since 2007 and was the founding president of IMSA from 1986 to 2007. IMSA cofounder and Nobel Laureate Leon Lederman was the former director of the Fermi National Accelerator Laboratory in Batavia. Dr. Marshall is a member of national and international advisory boards, commissions, and peer-review boards in innovation, design, STEM education, and research. She has been the keynote speaker for major national and international conferences and symposia. Dr. Marshall was chosen for Illinois's highest honor for achievement in 2005, the Order of Lincoln Laureate. She was also inducted into the Illinois Hall of Fame in 2007. Her husband, Robert, is a retired assistant superintendent from the Glenbard High School District. They have two children, Stacy and Scott, and five grandchildren.

CHAPTER TWO

Dancers and Singers, Artists All

Long ago, far away. Life was clear, close your eyes. Remember is a place from long ago. Remember filled with everything you know.

—Harry Nilsson, "Remember"

Wendell Minor
A 1962 graduate of West Aurora High School, Minor (b. 1944) is nationally known for the artwork he has created for more than 50 children's books. He received an honorary doctorate of humane letters from Aurora University in 2004 and from the University of Connecticut in 2010. His illustrations and design have enhanced over 2,000 works. In 1988, Minor was chosen as one of a six-member team commissioned by NASA to document the shuttle *Discovery*'s return to flight. He has created paintings for Chicago's Adler Planetarium and Astronomy Museum and for the US Postal Service. As president of the Society of Illustrators from 1989 to 1991, Minor considers one of his most important contributions the organization of an international exhibition and the subsequent publication of the book of the same name, titled "Art for Survival: The Illustrator and the Environment." He serves on the board of trustees of the Norman Rockwell Museum and is a member of the Low Illustration Committee at the New Britain Museum of American Art. He and his wife, Florence Friedmann Minor, have collaborated on many projects, including the recent tour of Pennsylvania's libraries, schools, Head Start facilities, and museums with their book *If You Were a Penguin*. Minor was named a Fox Valley Arts Hall of Fame artist in 2002.

Nancy Smith Hopp
Born to Charles Dudley and Margaret McWethy Smith in 1943, Nancy Hopp lives in her childhood home in West Aurora. She graduated from West Aurora High School in 1961, and earned a bachelor of arts degree in social sciences and a master of science degree in business management from Aurora University. Among her honors are the Aurora Kiwanis Club's God and Fellowman Award in 2010, the Lyle E. Oncken Community Service Award in 2004, the YWCA Woman of Distinction Award in 1990, Optimist of the Year in 1987, and the National Charlotte Danstrom Women in Management Achievement Award in 1984. Community involvements for the past 40 years have included the Paramount Arts Centre Endowment and the Exchange Club of Aurora. Hopp currently serves as a trustee for the Fox Valley Park District and as secretary for the Fox Valley Arts Hall of Fame. Hopp has written 36 short stories for her grandchildren and the first full-length, illustrated biography of artist Ruth Van Sickle Ford. *Warm Light, Cool Shadows: The Life and Art of Ruth Van Sickle Ford* was released after more than seven years of research. Nancy and James C. Hopp of Aurora wed on February 4, 1978. They have one son, Edward Thompson Reid Jr., and five grandchildren.

Maud Powell (RIGHT)
Powell was born in Peru, Illinois, on August 27, 1867, and grew up in Aurora. She was trained by the best American and European teachers and became America's first great master of the violin to achieve international rank. At the age of 16, she toured Great Britain, where she performed before the royal family. Powell was the first solo instrumentalist to record for the Victor Talking Machine Company's celebrity artist series, in 1904. She introduced chamber music with the Maud Powell String Quartet in 1894 and the Maud Powell Trio in 1908. The New York Symphony paid tribute to her as a "supreme and unforgettable artist." While on tour, Powell died of a heart attack on January 8, 1920. She is considered the first great American master of the violin, for her commanding bow and magnetic personality. Powell was a charter member of the Fox Valley Arts Hall of Fame, inducted in 2002. Her father, William B. Powell, was superintendent for East Aurora District No. 131. (Courtesy of Karen A. Shaffer and the Maud Powell Society for Music and Education.)

Paramount Arts Center

In 1978, the Paramount Theatre was returned to its original grandeur with a $1.5-million restoration. This 1931 Art Deco theater is listed in the National Register of Historic Places. It was designed by C.W. and George L. Rapp. In 2006, a 12,000-square-foot lobby, costing $6.2 million, was added. Tim Rater is the current executive director. Diana Martinez, who preceded him, is now president of the Second City in Chicago. (Author collection.)

Hollywood Casino Aurora

Hollywood Casino opened with a "Puttin' On the Ritz" theme on June 17, 1993, with a pair of riverboats. In June 2002, the riverboats were revamped and replaced with a 53,000-square-foot facility on one level. The casino includes 18-foot-high ceilings, 1,172 slot machines, 7 poker tables, and 21 gaming tables. There have been 1,498,609 visitors to date. (Courtesy of Jim Hopp.)

Richard Haussmann
Haussmann was born on March 22, 1930, in Aurora to George and Hazel Burrington Haussmann. He graduated in 1948 from East Aurora High School and in 1950, joined the Air Force. He began working at Old Second National Bank in 1957 as a teller and later handled the bank's marketing, advertising, and public relations duties. In 1982, he became the vault manager, serving in that position until his retirement in 1995. His was the voice for numerous band concerts and Aurora Historical Society events, and he served as Post No. 20, GAR Memorial Hall chaplain. Haussmann was a thespian who appeared in plays for the Aurora Drama Guild, the Boulder Hill Playhouse, the Albright Theatre in Batavia, and the Pheasant Run Playhouse in St. Charles. He received the Aurora Historical Society's Lifetime Achievement Award just prior to his death. He was able to watch the tribute video in his honor from his bed at Rush-Copley Medical Center in Aurora, where he died on March 26, 2001.

Ruth Van Sickle Ford
Born on August 8, 1897, in Aurora, Ruth Van Sickle Ford graduated from West High in 1915 and from the Chicago Academy of Fine Arts in 1918. She continued her studies in New York and Chicago under the tutelage of some of the most respected painters of the time, including John Carlson, George Bellows, and Guy Wiggins. That same year, 1918, she wed Albert G. "Sam" Ford; they were married for 60 years before his death in 1984. Ford's paintings were shown at the "Century of Progress" exhibition at the Art Institute of Chicago in 1933–1934. She became the academy's president and director, a post she held for 23 years, and was the first Illinois woman to be invited to join the American Watercolor Society in 1954. Ford was a founding member of the Chicago Women's Salon. Throughout her career, she was featured in one-woman shows in New York City, California, North Carolina, Mexico City, Haiti, and the Chicago Art Institute. In 1960, she became the first woman artist member of the Palette & Chisel Academy in Chicago and was the first American artist to exhibit in Haiti, at the island's national art museum, displaying her Caribbean watercolors. Aurora University awarded Ford an honorary doctor of fine arts degree in 1974. She died in 1989 and was inducted posthumously into the Fox Valley Arts Hall of Fame in 2002. Ruth Van Sickle Ford was survived by her daughter Barbara, who was a 1936 graduate of West Aurora High School.

Lucille R. and Sten G. Halfvarson

Lucille Robertson was born on May 17, 1919, the only child of Harris Morton and Lucille Fox Robertson, and grew up in Galva, Illinois. She received her bachelor of arts degree and graduated Phi Beta Kappa from Knox College in Galesburg, Illinois, in 1941. She took her master's degree of music from the American Conservatory in Chicago in 1969. Sten Halfvarson was born in Wilkinsburg, Pennsylvania, to Swedish immigrant parents in 1915. Lucille and Sten were married in Galva on August 8, 1946, and they settled in Aurora to raise their children, Laura, Eric, Linnea, and Mary. Eric became a renowned international opera singer. Lucille established the Waubonsee District Chorus, which became known for its annual performance of *The Messiah*, held at the Paramount Arts Center from 1968 to 1992. She was named YWCA Woman of the Year in 1976, a Paul Harris Fellow from the Rotary Foundation in 1999, and was the recipient of the Distinguished Service Award from the Cosmopolitan Club of Aurora in 1983. Sten was born in Wilkinsburg, Pennsylvania, in 1915. He was a choral director for more than 50 years, teaching at West High School from 1938 to 1979, and he directed the Illinois All-State Choir in 1968, which performed at the World Conference in Dijon, France. Aurora University awarded honorary doctoral degrees to Sten and Lucille in 2000. The Halfvarsons were survived by six grandchildren and one great-grandson. (Courtesy of the Community Foundation of the Fox River Valley.)

Arlene and Richard "Dick" Hawks

Arlene was born on June 27, 1949, in Newark, New Jersey, to Rose and Victor Marano. She earned her bachelor of arts degree in speech and theater from Adelphi University in Long Island, New York, and performed as a solo vocalist for Artist Entertainment Corporation, which brought her to Aurora. She and Dick were married on June 23, 1974. Their son Victor was born on April 30, 1975. Arlene retired from East Aurora District No. 131 in 2005 after 32 years of teaching. She was chosen the grand marshal of the Fourth of July parade in 2004 and was named a YWCA Woman of Distinction in 1988. She is a board member for the Aurora Public Art Commission, serves on the President's Advisory panel for CASA Kane County, and is a member of Kiwanis and Delta Delta Delta Alumni philanthropy. Arlene and Dick were honored by the Aurora Public Library Board in 2010. She now serves as director of the Fox Valley Park District's summer stage program. Arlene was awarded a Martin Luther King Jr. Award at West Aurora High School in January 2012 for community service to youth. Dick was born on July 11, 1939, to Catherine and William Hawks. He graduated from Marmion Military Academy and Aurora University. He has been a professional actor, a successful stockbroker with a trading seat on the Mercantile Exchange in Chicago, and has enjoyed playing polo. He has served on the board of the Aurora Civic Center Authority since 1987. Dick and Arlene have been named the new cochairmen of the Paramount Arts Centre Endowment Board.

Bruce and Claire Newton
Bruce (left) was born on December 21, 1926, and Claire (center) was born on August 21, 1928. Bruce had a tour with the Navy and came back to the Art Institute of Chicago to do postgraduate work. Claire graduated from West Aurora High School. Ruth Van Sickle Ford had given her a scholarship to the Chicago Academy of Fine Arts, and it is there that Bruce and Claire met. Bruce was on the staff of *The Tom Wallace Show* at WGN when Frazier Thomas suggested they begin a children's show, and the puppet Garfield Goose was born. Garfield was created in Aurora; a local man made the metal beak, and Claire sewed his button eyes. Another 270 puppets would be created by the pair. Bruce moved to ABC as staff producer, performer, and writer for the next 14 years before leaving to cofound WCIU, Channel 26. While he was working in Chicago, Claire was in Aurora, raising their five children: Sandra, Darry, Jim, Robin, and Tammy. They performed at the Union League Club Show at Christmas time and at the Winter Club in Highland Park for the Marshall Field family. Claire died on July 3, 2006, and Bruce passed away on November 8, 2007. As William Carlos Williams wrote in *Spring and All No. XXI*: "So much depends upon / a red wheel barrow / glazed with rain water / beside the white chickens." This photograph is from a 2000 "Millennium Moments" event with the author (right).

Susan Unger Wyeth

Wyeth was born in Rogers City, Michigan, on June 8, 1959, to Ralph and Barbara Unger. She has two older brothers, Bob and John, and two younger sisters, Vicki and Sarah. She received her bachelor of arts degree from Michigan State University and worked at a local cable station. In 1982, she met Jay Wyeth at Concordia College in River Forest, and they dated while he attended law school at the University of Illinois. They married on October 11, 1986, and are the parents of four sons: Ryan, Jason, Austin, and Brandon. Susan teaches second grade at St. Paul's Lutheran School. She is the author of *There's a Flood in My House*, a book that describes the 1996 Aurora flood and her boys' reaction to it. On July 17 and 18, 1996, a record amount of rainfall in a 24-hour period (16.9 inches) inundated the city. The book is full of humor and whimsical illustrations of cut-paper designs by Susan. She has served on the board of directors for the Lutheran Camp Association and was the part-time children's ministry director for St. Paul's.

The Hawthorne Book Club

The Hawthorne Book Club started unofficially in 1885, when friends of Herminone (R.W.) Gates read to her during an illness. In 1886, members began to visit on the same day, Monday, and to take up a course of study. In 1893, a name was selected and officers were chosen. Gates lived on Hawthorne Court, so that name was chosen in her honor. Aurora had become known as the "City of Lights" after the erection of its light towers in the 1880s. The oldest club yearbook is from the 1894–1895 season. At that time, the club had met for 25 afternoons. During that season, the members studied Ireland, Scotland, and Wales. For many years, research papers were prepared for the programs rather than reports on specific books. Current officers of the club are president Lois Swan, vice president/programs Irma Larson, secretary Nita Hayward, treasurer Geri Galli, and historian Jane Zimmerman. (Courtesy of the Aurora Historical Society archives.)

1895
Social butterflies.
Aren't they sweet?

Marissa Amoni
Born on August 20, 1976, to Jill Kay Rogers and Daniel Joseph Amoni of Aurora, Marissa Amoni graduated from West Aurora High School in 1994, and graduated with a double major in philosophy and psychology from the University of Arizona. Amoni married Max Landon Balding on May 26, 2010, and they are raising two children, Stella and Guy. Amoni is the founder and editor of *Downtown Auroran* magazine, the organizer of Alley Art Festival in downtown Aurora, and she initiated Downtown Aurora Arts Mixers (DAAM).

Fox Valley Symphony
The Fox Valley Symphony began as a chamber orchestra in the spring of 1958, with the first concert performed on December 4, 1958. The first conductor was Dudley Powers, former first cello of the Chicago Symphony Orchestra. The symphony ceased performing in 2001 as a result of declining subscriptions following the 9/11 terrorist attacks. The Fox Valley Philharmonic Symphony Orchestra began in 2010 under the direction of Dr. Colin Holman. In this photograph are officers Shirley B. (Richard) Stolley and Joseph Holty. (Courtesy of Joyce Dlugopolski.)

Joy Tarble

Tarble, born in New Hampshire on April 27, 1806, wed Harriet Cox on November 26, 1829, in Adams, New York. They arrived in Aurora in the spring of 1846. Their four children were Melvin, Martin, Matilda, and Myron, and the couple had eight grandchildren. In November 1879, they celebrated their 50th wedding anniversary. They greeted friends at their home at 523 Lake Street for four hours; the presents they received were numerous, including $200 in gold coins. Tarble was a staunch Republican. He was street commissioner of the West Division, earning $1.25 daily, and was sexton of the West Aurora Cemetery. In the 1859 Aurora City Directory, he is listed as "mason, north part of city." He built the foundation of the Tanner house, homes for L.D. Brady and the Honorable B.F. Fridley, the "big stone wagon shop" at the west end of the bridge, his home at 453 Plum Street (pictured), and many brick and stone dwellings. Grandson Albert Tarble, captain of the Aurora Zouves, lived directly behind him. The Zouves was a world-famous drill team c. 1900. Joy Tarble died on August 19, 1889. His great-great-granddaughter is Maureen Avery Granger, who was born on November 6, 1929, to Roland and Ireane Boldin Avery. Maureen and Robert Granger wed on May 14, 1949, and had four children: John, Jeff, Michael, and Martha. Maureen obtained her bachelor and master arts degrees from NIU, DeKalb as well as a certificate of advanced study in educational administration in 1984. She taught with West Aurora Schools for 12 years before becoming principal at McCleery School for 16 years. She retired in 2000. Maureen has six grandchildren. She is the current president of the Aurora Golden K Kiwanis.

Delbert Peterson
Peterson was born on June 10, 1926, in Aurora and lived at 210 South Union Street. He graduated from East Aurora High School in 1944 and received a scholarship to the Chicago Academy of Art before serving in the Army Air Force during World War II. Peterson worked as an artist at the Geneva Kitchens and at Elgin Watch before opening his Batavia art studio in 1955. He married Jeanne Mae Ruther in 1949, and they lived in Batavia for 50 years, raising their seven children before moving to Vero Beach, Florida. Peterson has exhibited across the United States, and his paintings are in the permanent collections of the Artists of America Calendars, 1980 and 1993 editions, as well as at the Metropolitan Life Insurance Company and Amoco locations. Peterson is a member of the Florida Watercolor Society and the Aurora and Naperville Art Leagues. He and his wife received the Laurel Award for Artistic Leadership in 2010 from the Cultural Council of Indian River County. He was chosen as a Fox Valley Arts Hall of Fame artist, class of 2012.

Charles P. Burton
Born on March 7, 1862, in Anderson, Indiana, Charles Pierce Burton arrived in Aurora at age 12. Charles was a historical editor and writer of children's books. His father, Pierce Burton, was a well-regarded newspaper publisher. The first Bob's Hill book was published by Henry Holt & Co. in 1905. Ever the humorist, Burton delighted in telling people that he was "the brightest boy in the 1880 graduating class at East High," only to admit that he was the only boy in the class. His wife, Cora Vreeland Burton, and two sons and a daughter survived him when death called on March 31, 1947. (Courtesy of the Aurora Historical Society archives.)

Tuesday Garden Club
The first garden club on record in Aurora was founded in 1930 by a group of men. They undertook the development of the rose garden at Phillips Park and other projects, including the community Christmas tree. The Tuesday Garden Club was founded in March 1936. Records begin with a September 13, 1938, meeting at which women were identified by their husbands' first initials and last name. The club brought home a purple ribbon for best of show from a 1947 event at Marshall Field's in Chicago. (Courtesy of Mary Clark Ormond.)

Mid-West Early American Pressed Glass Club
In 1933, four Aurora women, Mrs. Justus Johnson, Mrs. H.D. Hallett, Mrs. Carl Grometer, and Mrs. William Berry, founded the Mid-West Early American Pressed Glass Club. Johnson had a collection significant enough that 200 pieces were displayed at the Chicago Art Institute, and a collection of her glass was accepted for permanent display at the Victoria and Albert Museum, London. Original dues were $2. For years, club members auctioned off their own antiques during meetings as a means of fundraising. Current leaders include, from left to right, (first row) Mary Blohm (first vice president), Jean Goehlen (president), and Maxine Jacobson (second vice president); (second row) Martha Egeland (assistant first vice president), Jane Zimmerman (secretary), and Sue Ellen Blazek (treasurer). Directors, not shown, are Laura Fulton, Faith Jones, Helen Thelin, and Dolores Palmquist. (Author's collection.)

Joyce Reuland (LEFT)
Born on August 21, 1940, to Ralph and Bernice Kunold of Aurora, Joyce Reuland graduated from St. Mary's grade school, then East Aurora High School in 1958. She wed Robert Reuland in 1964, and they had three children, Scott, Steven, and Mary Beth, and four grandchildren. Reuland studied with Ruth Van Sickle Ford and Delbert Peterson. In 1999, she opened Gallery 44, a cooperative of area artists. She is a member of the Aurora Public Art Commission and has exhibited her work at the Anam Art Gallery, the Waterfront Gallery at the Abbey Resort, and exhibited in juried exhibits throughout the Midwest. Reuland has done hundreds of commissioned works throughout the country. She is enjoying a new style she calls "impressionistic abstract with mixed media." She completed some pen and ink drawings for the Dominican Literacy Center's garden walk and has exhibited in the Grand Gallery of the Paramount Arts Centre in 2007, 2008, and 2009. Reuland is using birch bark in collages and feels that its natural beauty and unique designs lend to interesting artwork.

Donnell Collins
Donnell Collins titled this photograph "Outstanding in its Field," and it epitomizes his excellent eye. Collins was born in Mississippi to Robert and Lurene Collins in 1955. He graduated from Oak Park Elementary, Simmons Junior High, and East Aurora High School in Aurora. His associate's degree is from Waubonsee Community College. He received his bachelor's degree from NIU, DeKalb. He served in the US Army from 1974 to 1977. Collins has been a photojournalist for the *Beacon News* for 20 years and has taken first place in contests sponsored by the Associated Press, the Illinois Press Association, and the Northern Illinois Newspaper Association. He has worked for magazines and newspapers, including the *Chicago Sun-Times*, the *Chicago Tribune*, and *ESPN The Magazine*. In 2004, he was honored by the City of Aurora African-American Heritage Advisory Board for his outstanding achievements. In October 2007, he was appointed to the Kane County Board and subsequently was elected to the position. Collins and his wife, Pam, attend Main Baptist Church and have been married for 34 years. They have three grown children, Jason, Shaun, and Kayla, and two grandchildren, Tyler and Duane.

CHAPTER THREE

A Wheelbarrow Full of Hope

There are only two ways to live your life.
One is that nothing is a miracle and the other is that everything is a miracle.

—Albert Einstein

Bronislaw (Bruno) Bartoszek

Bruno Bartoszek (1922–2010) was born in Toporow, eastern Poland. He served with the 3rd Division of the American Army and saw action at Monte Cassino in 1944. He was bestowed with Poland's highest military honor, the Virtuti Militari, for his wartime service. After he arrived in Aurora in 1951, he became an American citizen. He worked for the Burlington Railroad and Aurora Pump, and he was with Prudential Insurance for 22 years. Bartoszek was chosen to be the grand marshal of Aurora's Fourth of July parade in 1996 and was selected Tutor of the Year by Literacy Volunteers of America–Illinois in Chicago. He was chosen as the Hal Beebee Book of Golden Deeds Award recipient in 1999. As a result of that honor, Mayor David L. Stover proclaimed April 8, 1999, Bruno Bartoszek Day in Aurora. He was chairman of the Kedeka District Food Drive, served on the religious awards committee of the Rockford Diocese, and was a counselor on the Citizenship Committee for Community, Nation and World with the Boy Scouts. Bartoszek was an adult literacy volunteer tutor for 20 years at Waubonsee Community College, teaching English skills to adults. On November 1, 2010, All Saint's Day, he passed on, leaving to mourn him son Jack, daughter Annette, son-in-law Rick, daughter-in-law Donita, and three grandchildren, Stephanie, Sarah, and Allen. He shall be remembered as a compassionate contributor to many in the Aurora area. Bruno was the gold standard of integrity.

Marie LeBeau Wilkinson
Wilkinson was born on May 6, 1909, in the French Quarter of New Orleans. She was raised a devout Catholic and studied business at the former Straight University (now Dillard University). She visited Chicago at age 20 and met her future husband, Charles, on a blind date. They married in 1930. They had two children, Sheila Scott-Wilkinson and Donald Wilkinson. Marie helped launch many charitable organizations, including the Marie Wilkinson Foundation and Food Pantry, the Marie Wilkinson Child Development Center, Hesed House Homeless Shelter, SciTech Youth Science Museum, and the local chapter of the Urban League (Quad County). In 1948, when she and her friend Bernice Christmas were refused seating at Hart's Drive-In in Aurora because she was of the "so-called colored African race," she took her case before the state appellate court (case No. 10361). Although the restaurant was not held accountable, an employee, Norbert Finney, was fined $25. Through the Human Relations Commission that she founded in 1964, she is credited with the first Fair Housing Ordinance in Illinois. Wilkinson served for over 30 years on the Aurora Human Relations Commission. In 2001, she received the Lumen Christi Award, the Catholic Church's highest honor for a missionary. She was the first nonordained person to receive the award. When she turned 100 in May 2009, the city of Aurora celebrated her life with a public birthday event. Wilkinson died a peaceful death on August 12, 2010. A statue of a seated Marie Wilkinson has been placed in her honor in front of the Aurora Public Library on Benton Street, where visitors can sit on the bench beside her.

Emily Fuller-Gibson
Fuller-Gibson was born on September 27, 1939, in Litchfield, Illinois, to Edward and Mildred Fuller. She wed Melvin Gibson on June 14, 1957, and they had daughters Michelle, Melanie, and Margot. Fuller-Gibson graduated from East Aurora High School in 1957. She graduated magna cum laude from California State University in 1981 and received her master of arts degree in English in 1982. She was the NAACP branch president and led the first sit-in to open public accommodations in Aurora, in 1966. Fuller-Gibson led the first Illinois civil rights demonstrations in Kane County for open housing also in 1966. (Courtesy of Malvin Stuart.)

Chief of Police Gregory S. Thomas
Gregory Thomas holds a bachelor's degree in criminal justice from Lewis University and an master of science degree in business administration from Aurora University. He began his police career as a cadet in 1978 and became a sworn officer in 1982. He was appointed chief on April 22, 2008. Thomas's numerous awards include the Kendall County Medal of Valor and the Exchange Club of Aurora Police Officer of the Year. He was born on September 26, 1960, and is a 1978 graduate of East Aurora High School.

Faith and Lloyd Jones
Faith Schule was born in Aurora on October 6, 1930, to Mack and Estelle Schule. Lloyd Jones was born on April 26, 1925, to Lloyd Sr. and Edith Jones. Faith received her degree in zoology from North Central College as preparation for her nursing career. Lloyd took his master's degree in music education from Northwestern and taught for 25 years in the West Aurora schools. They were married on October 2, 1954, and raised three children, Bradley, Laurel, and Alan.

Warren Humiston

Humiston was born on June 25, 1931, and graduated from East Aurora High School in 1951. He began with the Aurora Fire Department on December 15, 1957, and retired on March 5, 1988. He married Ann Powers on April 27, 1952. They had five children, Kathleen, Karen, Bill, Kristine, and Allan, and have 10 grandchildren and 10 great-grandchildren. Warren Humiston died in 2003. Ann has worked at Schaefer's Greenhouse for the past 33 years, expediting floral deliveries.

Laurie Jean Davis

The first 911 system installed for the city went online in the fall of 1978. For the past 22 years, when the 911 telephone CAD (Computer Aided Dispatch) console has rung at the police headquarters in Aurora, Laurie Davis has answered the call. Her colleagues include Ginger Chione, a 911 operator for 36 years; Melissa Johnson, 25 years; and Kelley Gruca, 21 years. In 2010, there were 274,000 calls received, and in 2011, a total of 258,000 calls were handled by the 30 operators. Davis was born on February 1, 1966, in Aurora to Larry and Beth Sehie. She began with the telecommunications section on October 9, 1989. Davis has taken many courses in leadership and supervisory skills and holds an Emergency Medical Dispatch Certification. She was promoted to a lead operator on January 1, 2003. Davis has volunteered with the Fox Valley Crisis Center and is a member of the Aurora Junior Woman's Club. Her children are Ryan and Rachel. (Author's collection.)

Aurora Police Department
The first marshal for Aurora, in 1857, was Dennis Baker, who was assisted by two constables. There were laws forbidding piano playing before 6 a.m. and riding or driving horses across the bridge faster than a walk. A gentleman who winked at a lady was subject to a 50¢ fine if the lady chose to complain. A police patrol wagon was maintained until 1911. The police department was located on East Downer Place from 1866 to 1966. On June 30, 1966, the Aurora Police Department opened at 350 North River Street, with capacity for 83 police officers and 32 civilian employees. This building closed in 2011 when the new, state-of-the-art police headquarters was opened on Indian Trail. There are currently 285 sworn officers, including 32 women, on the Aurora police force. Shown here is a 1917 program cover for the Aurora Policeman's Benevolent Association Dance. (Courtesy of Detective Lee Catavu.)

Gary and Mary Ruth Roberts
Gary Roberts was born on December 21, 1927, to Roy and Rhea Garrison Roberts in Marion, Illinois. The family moved to Aurora in 1930. Gary graduated from East Aurora High School and then attended Army Officer Candidate School (OCS), serving in Japan. Upon his return, Gary worked for several coal companies, becoming vice president of the Freeman United Coal Company in Chicago. Mary Ruth Farrell was born on December 19, 1928, to Edward and Ruthanne Burnett Farrell. She graduated with her nursing degree from St. Francis in Evanston and earned her bachelor degree in nursing at Loyola in 1951. Gary and Mary were the first couple wed at the current Holy Angels Church, on February 9, 1952. They are the parents of five children: Jeffrey, Beth, John, Ellen, and Patrick. They celebrate with 11 grandchildren and three great-grandchildren. They have volunteered for decades for the Holy Angels Food Bank, Meals on Wheels, PADS, the Mercy Hospital Auxiliary, the Mercy Hospital Board, and the Adult Literacy Project at Waubonsee Community College. Mary Ruth was a board member of the Community Foundation of the Fox River Valley from 1986 to 1998. She is a breast cancer survivor and volunteered for 30 years with the American Cancer Society. (Courtesy of Photography by Feltes.)

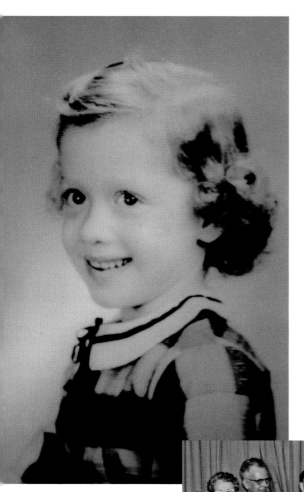

Marissa Happ

Happ earned her bachelor of science degree in education/childhood development from NIU, DeKalb, and her master's degree in social work from Aurora University in 2000. She and Bob Happ wed on October 4, 1975, and they are the parents of five grown children and have two grandchildren. Marissa has taught early childhood classes at Waubonsee Community College for the past 20 years. She has been a teacher and social worker in the behavior health area of Mercy Center.

Cosmopolitan Club

The Cosmopolitan Club began in Aurora in 1927 with an inaugural banquet on June 6, held at the Masonic Temple. Frank Weisgerber was the first president. The club's main focus was diabetes education, wellness, and research for a cure. The club provided scholarships for diabetic children to attend summer camp and funds to support the local Red Cross, Community Chest, Wayside Cross Mission, and the Polio Fund. This photograph shows the member's convention held at the Pere Marquette Hotel in Peoria in 1958. (Courtesy of Kenlyn Nash-Demeter.)

Jim Shazer

Shazer (b. 1927) was born in Uniontown, Pennsylvania, and arrived with his family in Aurora in 1953. He worked for Richards-Wilcox for 35 years. Shazer became a literacy volunteer with Waubonsee Community College in 1989 and was chosen as national Tutor of the Year by Literacy Volunteers of America (LVA) in 1997. At that time, LVA had 393 affiliates in 44 states, so this was an august honor, indeed. Shazer lived with his sister Marge in Aurora. He wed Donna Crabbe in 2003, and she passed in 2008. Jim was an avid gardener and today lives with family in Oglesby, Illinois. (Author's collection.)

Diane Picciuolo

Picciuolo was born in San Antonio, Texas, to Margaret and Jules Picciuolo in 1954. The Project Unity movement in Aurora helped to provide community services to improve conditions for underprivileged children and their families. Under that umbrella, Picciuolo originated the Back to School Fair in 1995, in which 50 community agencies and organizations participated. Today, the Back to School Fair serves 3,000 children. In 1998, she wrote the winning proposal for the Illinois Governor's Home Town Award for the City of Aurora/ Back to School Fair. Picciuolo has one son, Christopher, and one grandson, Darren.

John Peter Galles Jr.

Galles, born on April 5, 1925, to John and Anna Galles, graduated from St. Therese grade school and East Aurora High School in 1942. He joined the Navy and served for three years on the Island of Guam. Upon returning home to Aurora, he began working at the Aurora Cotton Mills. He married Emilie Rippinger on Thanksgiving Day 1947, and they had five children: David, Mike, Terry, John, and Josephine. They have 13 grandchildren and five great-grandchildren. Galles worked at Drake Surplus, handling war surplus equipment from the War Assets Administration until he opened his own store in 1982 on Downer Place. Galles is a member of the East Aurora Sports Booster Hall of Fame, the Fall Basketball League, the Aurora Superstars Youth Football League, the Aurora Bowling Hall of Fame, the Aurora Boy's Baseball Hall of Fame, and the Aurora Boy's Baseball Board of Directors. He has received numerous community honors, including being named the grand marshal of the Fourth of July parade in 1989, being given the keys to the city of Aurora at a city youth banquet, and having Solfisburg Park renamed the John P. Galles Park Baseball Complex in 1989 by Aurora mayor David L. Stover. On April 28, 2011, John received the Community Service Award from the East Aurora Alumni Association for his outstanding community service.

Charleen and Paul Reuland
Charleen and Paul Reuland founded Reuland Food Service in 1955. They were wed on June 21, 1944, and their children are Paul, Dan, Jeff, and Rene. Charleen's recipes continue to be favorites, including her potato salad and hot chicken salad. She passed on June 6, 2011, at age 86, and Paul died on March 26, 2000. Their grandchildren include Brigit Reuland-Foster and Ryan Reuland, who continue to staff the popular restaurant with their dad, Jeff.

Joseph Henning
Born on October 29, 1966, in Sycamore, Illinois, Henning graduated from NIU, DeKalb. He accepted a position with the Paramount Arts Theatre in 1989. Since 2005, Henning has been president and chief executive officer of the Aurora Regional Chamber of Commerce. During his tenure, the chamber has been recognized by state and national chambers for its work on behalf of the business community.

Clarence and Dorothy Culkin Ruddy

The Ruddy family arrived in Aurora in 1882, and that same year, James J. Ruddy started a plumbing business. Clarence J. Ruddy (top) was born in Batavia on June 20, 1905. He moved with his parents, Clarence S. and Mae Behan Ruddy, to Aurora, graduating from Our Lady of Good Counsel grade school. In 1922, he graduated from East Aurora High School. With his red hair, he earned the nickname "Torchie." At Notre Dame, he knew Knute Rockne, who called him "Red." He received his undergraduate and law degrees with high honors from the University of Notre Dame, where he founded and became the first editor-in-chief of the *Notre Dame Lawyer*, now known as the *Notre Dame Law Review*. In 1927, he was an associate in the law firm of Alschuler, Putnam, Johnson and Ruddy before opening his own firm in 1951 with partner Charles J. Myler. The firm is now known as Myler, Ruddy & McTavish. He was appointed the Illinois assistant attorney general on Valentine's Day 1933. Ruddy was president of the Aurora War Fund in 1943–1944, a member of Holy Angels parish, and was active in the Loyal Order of Moose. He was awarded the Pilgrim Degree of Merit, the order's highest degree, in 1942. He served as the fraternity's general counsel for the Supreme Lodge for over 20 years. In 1970, Ruddy received the Distinguished Service Award from the Aurora Jaycees. Ruddy's sterling character was a hallmark of his career. He married Dorothy Culkin (b. 1911), pictured at left, daughter of Louis and Cecilia Cosgrove Culkin, on April 20, 1940. On September 9, 1941, their son Philip was born. Daughter Mary Margaret and son Clarence John Jr. followed. Dorothy served on the Aurora Library Board from 1962 to 1982. Clarence's death date was June 21, 2004. His wife of 57 years, Dorothy, had preceded him on February 7, 1997. They had seven grandchildren.

Colleen and Philip Ruddy

Colleen Dee Murray was born on November 19, 1942, in Glendale, California, to Lyle "L.A." and Delores Murray of Minneapolis, Minnesota, and Key Biscayne, Florida. She graduated from St. Margaret's Academy in Minneapolis in 1960. She married Philip Ruddy on August 14, 1965, and they have two children, Erin and Philip Jr., and three grandchildren, Benjamin, Julia, and Reed. Colleen received her undergraduate degree from St. Mary's College in Notre Dame and her master of business administration degree from Aurora University. She was director of the Management Center at Aurora University for 10 years before becoming president of Valley Industrial Association in 1992. She retired on May 30, 2009. For the past 22 years, Colleen has managed cottage rentals in Harbor Country, Michigan. Philip C. Ruddy was born on September 9, 1941, to Clarence and Dorothy Ruddy in Aurora. He graduated from Holy Angels in 1955 and from Marmion Military Academy in 1959. After graduation from Notre Dame Law School in 1966, he was selected to serve as a law clerk to the chief justice of the Illinois Supreme Court, Roy J. Solfisburg. Philip was vice chairman of the Aurora Human Relations Commission and chair of the Holy Angels parish council, having previously served as president of the Holy Angels School Board. In 1970, he was chosen by the Aurora Junior Chamber of Commerce as Aurora's outstanding young man and subsequently chosen by the Illinois Junior Chamber of Commerce as one of the three outstanding young men in Illinois. He also served as a special assistant attorney general from 1984 to 1988 and was founder and original director of the Kane County Bar Foundation from 1996 to 2001. Philip was appointed corporation counsel for the City of Aurora by Mayor Albert D. McCoy in May 1973. He has been associated with his father's law firm since 1968, which is now known as Myler, Ruddy & McTavish.

Lorna and John Ruddy

Here is a family photograph of the John and Lorna Ruddy family taken on the Inca Trail, Machu Picchu, on October 30, 2009. Shown are, from left to right, (first row) two unidentified people; (second row) Lorna (b. 1951), Alison, Jay, Erica, and John Ruddy (b. 1950). John is the younger son of Clarence and Dorothy Ruddy. John received his juris doctorate from John Marshall Law School in 1977 and his bachelor of arts degree from the University of Denver. He is an attorney with the firm of Ruddy, Milroy & King. Lorna works at Provena Mercy Center as a medical technologist.

Sherman L. Jenkins

Jenkins was born on September 27, 1956, in Chicago, the son of Marie Jenkins. He graduated from NIU, DeKalb, with a bachelor of arts degree in journalism. He is the executive director of the Aurora Economic Development Commission. He is a member of the Rotary Club and a trustee of the Copley Healthcare Foundation at Rush-Copley Medical Center and is past chairman and board emeritus of the Aurora Public Library Foundation. Jenkins married Juliette Munnerlyn on August 6, 2006. They have two children, Knasi and Ayanna, and one grandson, Jai Antonio.

Mary Jane Tharp Hollis
Mary Jane Tharp was born in
Hillsboro, Ohio, to Dr. Vernon and
Grace Tharp in 1942. In 1963, she
came to Aurora and began a children's
TV show on WLXT. She started the
Aurora Community Study Circles in
1996 after receiving her bachelor of
arts degree from Aurora University
and her master of arts degree from
NIU, DeKalb. The Aurora Community
Study Circles collaborated with 30
area organizations, including the
City of Aurora, to promote public
education and to increase civic
understanding and participation in
interracial problem-solving and bias
reduction. "Many Young Voices"
began at West High, culminating
in the student formation of a
multicultural club and the first Mosaic
of Youth Conference for area high
schools. Hollis has been a member
of the Aurora Junior Woman's Club,
Children's Dental Board and Dental
Ball (chairperson), board of directors
of Child Welfare Society at Jack and Jill
Nursery School, Copley Hospital Gift
Corner (cochairperson), Fox Valley
Symphony Board (vice president),
YWCA (president), and has served
on the Ice Skating Institute of America
Board of Directors. Hollis was named
a YWCA Woman of Distinction in
1998, received the Peace Leader
Award in 1999 from the Illinois Center
for Violence Prevention, and the
2002 Illinois American Association of
University Women, Agent of Change
award. She has four children: Ellen,
Amy, and twins Michael and Claire.
Hollis has 10 grandchildren.

Mary Clark and Neal Ormond III

Mary Clark was born on November 24, 1946, in Chicago to Lawrence and Albina Urba Clark. She received her bachelor's degree in English/music in 1968, her master of library science degree in 1969 from Rosary College, and her master of business administration degree from Loyola University in 1975. Mary was head librarian at the Aurora Public Library from 1975 to 1979. She wed Neal Ormond III on May 27, 1978, and they have three children: Neal Alan, Laurel, and Chrissy. Mary has served on many boards, including as first president of the Fox Valley Arts Hall of Fame Board founding member and served on the Aurora Public Library Foundation Board. She is board president of the Aurora Historical Society. She was honorary chair, with Neal, for the Aurora Public Library annual gala, and the grand marshal, with Neal, of the Independence Day parade in 2010. She has designed the Environmental Sciences Courtyard at West High. Mary was chosen a YWCA Woman of Distinction in 1979. Neal's great-great grandfather emigrated from Ireland in the 1860s. He was a Yonkers, New York, policeman. Neal Ormond III was born in New York on February 16, 1943, to Neal and Margery Ormond. He graduated from West Aurora High School in 1958, earned his bachelor of science degree in engineering at Yale University in 1962, and earned his master of business administration degree from Stanford University graduate school of business. He attended John Marshall Law School in 1964–1965. Neal has performed radio and television sportscaster duties, including 47 years as the "Voice of the West Aurora Blackhawks." He is cofounder, past president, and board member of the Blackhawk Sports Boosters Club and has served on the District No. 129 Board of Education for the past 17 years. Neal is chairman of the Community Foundation of the Fox River Valley. He was installed in the Illinois Basketball Coaches Hall of Fame in 1984 and in the West Aurora Sports Hall of Fame in 1989. He was homecoming parade grand marshal for West Aurora in 2003, and the City of Aurora proclaimed Neal Ormond Day on February 6, 2004.

The Huntoon Family
Philip Huntoon came to America in 1688 from the county of Wiltshire in England. The Joseph Huntoon family settled in the Aurora area two years before Aurora became a city. In the mid-1860s, E.D. Huntoon rebuilt a hotel that had burned down, and it became known as the Huntoon House. At that time, Galena Boulevard was a dirt street on which stagecoaches and teams rode past. The building still stands at the southeast corner of Galena Boulevard and Middle Avenue. Howard bought a horse, which he rented out and with which he also gave riding lessons. This was the beginning of the Huntoon Stables. The ninth generation of this family in Aurora includes Carleton Huntoon, who was born on January 30, 1940, to Howard and Margaret Huntoon on the family farm. He wed Melinda Mann, and they have one son, Christopher. Carleton's parents purchased the 270-acre farm property in Sugar Grove in 1939 for $94 an acre. About 186 acres were sold in 1967, at $1,800 an acre, for the Waubonsee Community College Sugar Grove campus. Carleton and his family still operate the stables on Oak Street in North Aurora. Also of the ninth generation is Nancy Huntoon Garbe, an avid horsewoman, wife, and mother. Shown in the photograph is Fred Huntoon c. 1890.

Chief of Police Robert E. Brent

The sheriff delivered Robert E. Brent on February 20, 1939, in Thackeray, Illinois, to Howard Earl and Mary Lorene Clauss Brent. Bob Brent went to Center Grade School and graduated from East Aurora High School in 1957. He attended the University of Illinois and Waubonsee Community College, the University of Louisville for the School of Police Administration, and the LaSalle Extension University for his bachelor of law degree in 1975, as well as many other police trainings and seminars. He wed Janice Rausch on July 9, 1960, and they have two children, Mark and Ann Marie. Brent became a police officer on August 14, 1961, and rose continually in the ranks until 1977, when he was appointed chief of police on September 1. He served for 12 years, until his retirement in 1989. Chief Bob has been an adjunct professor at Aurora University in its criminal justice management master's degree program. He worked for Alarm Detection Systems as director of public relations from 1989 to 2004. He was named Police Officer of the Year by *Police Magazine* in 1968, was Kiwanian of the Year in 1985, and was club founder of the Golden K Kiwanis Club. He received the American Legion God and Country Award in 1953. Brent was grand marshal for the Fourth of July parade in 2007. In 2012, he was recognized as the 1985 founder of Crime Stoppers. (Courtesy of Photography by Feltes.)

CHAPTER FOUR

Serving with Distinction

Doesn't everything die at last, and too soon?
Tell me, what is it you plan to do with your one wild and precious life?

—Mary Oliver

Mayor Albert Denis McCoy

Albert McCoy was born in Aurora on June 25, 1926, to Denis and Katherine McCoy. He graduated from Marmion in 1944 and joined the Navy, earning five battle stars. He wed Mary Ann Malmborg on November 25, 1954, and they had two children, Cara and Michael, and five grandchildren. After the war, McCoy attended Aurora College and the University of Montana in Missoula on a football scholarship. In 1965, at the age of 38, he was elected mayor of Aurora, the first Marmion graduate to hold that office. He served from 1965 to 1977. During McCoy's tenure as mayor, Aurora annexed 6,800 acres of land in DuPage County, the second-largest annexation in Illinois history. The transaction included the Fox Valley Villages area and the Fox Valley Mall. The main east-west street to the mall was later named McCoy Drive. Among his many accomplishments were the reorganization of the Aurora Police Department, building new central police and fire stations, and obtaining the 1976 federal aviation control tower for Aurora airport in Sugar Grove. It was his brilliance that saw Aurora pass a fair housing ordinance, the second city in Illinois to do so. He saved the Paramount Theatre from demolition in 1978, turning it into the Performing Arts Center. McCoy was the only mayor in western suburban cities who voted to form the Regional Transportation Authority. He served as chairman of the Illinois Liquor Control Commission from 1977 to 1998. He served on the Marmion Alumni Board of Directors for nine years, the Serra Club for 26 years, and the Lions Club for 46 years. McCoy was a lector at Holy Angels Church for 27 years. He is a recipient of the Marmion Centurion Award, the Ad Regnum Dei, and another 30 awards for his capable and astute leadership. His wife, Mary Ann, died on June 27, 2008.

Mike McCoy

Mike McCoy was born to Al and Mary Ann McCoy on June 2, 1956. He is a 1974 graduate of West Aurora High School and a 1978 honors graduate from the University of Illinois with a bachelor of science degree in civil engineering. Mike is a senior project manager at Omega & Associates in Lisle, Illinois, where he manages construction engineering for large transportation projects, including the resurfacing of the Eisenhower and Stevenson Expressways. He was chair of the Kane County board from 1996 to 2004, and served as chair of the Kane County Transportation Committee and the Kane County Board of Public Health. He is past president of the Metro Counties Association and chair of the Aurora Board of Election Commissioners. McCoy was appointed to the Commuter Rail Board (Metra) in 2011 and has received the Lifetime Achievement Award from the Quad County Urban League, as well as numerous awards recognizing his efforts in land preservation and engineering. Mike is married to Victoria Leonardi, and their children are Joseph, Kate, and Mark.

Aurora City Council 2012

Shown here are, from left to right (first row) Robert O'Connor, Mayor Thomas J. Weisner, and Richard C. Irvin; (second row) Abby Schuler, Juany Garza, Stephanie Kifowit, Richard Lawrence, John S. "Whitey" Peters, Michael B. Saville, Scheketa Hart-Burns, Richard B. Mervine, Allan Lewandowski, and Lynda Elmore.

Mayor Thomas J. and Marilyn Weisner

Tom Wiesner was born in Batavia to Cassie and Jack Weisner on October 6, 1949. He is a graduate of Marmion Academy and Aurora University, summa cum laude. He and Marilyn were married on August 12, 1972. Marilyn was born on November 13, 1950, to Margaret and Frank Hogan. Tom served with Marilyn in the Peace Corps on the Solomon Islands. She is the current director of Aurora Area Interfaith Food Pantry. Their son Thaddeus (1984–2006) was born on the island of Guadalcanal. Their foster son is Anthony Scotti. Tom is founding chair of the Aurora East Education Foundation and cofounder of Aurora Care Corporation. Tom was first elected mayor of Aurora in 2005.

Aurora Public Library Board of Directors
The Aurora Public Library was established by city ordinance in 1881 and was located in the Grand Army of the Republic Memorial Hall at 23 East Downer Place until 1904. Using a $50,000 grant from philanthropist Andrew Carnegie, a new building was built at 1 East Benton Street. In 1969, the library underwent a major renovation, when two three-story wings and a modern facade were added, expanding the building to its current 44,000 square feet. In 1993, the Eola Road Branch Library opened, and in 1998 the West Branch Library opened. The current library board is planning ground-breaking for a new, state-of-the-art main library. The library board members, shown here from left to right, are (first row) Anthony Stanford, Dr. Jill Wold, head librarian Eva Luckinbill, and board president Jeffrey A. Butler; (second row) board secretary Norma Gobert, John Savage, Dick Hawks, and Walter Meinert. Not shown are board members Norma Vazquez and Jeffrey Redding. (Courtesy of Dave Chrestenson.)

Eleanor Plain
Plain (1905–1981) was born on September 15, 1905, the daughter of Judge Frank G. and Jennie Guinong Plain. She graduated from the University of Michigan and earned her master's degree from the University of Chicago. She received an honorary doctorate from Aurora College in 1976. Plain joined the Aurora Public Library staff in 1931 and was head librarian from 1939 to 1975. She was a past president and member of the board of directors of the Aurora Historical Society. She was listed in five *Who's Who* reference books and was an accomplished pianist and organist. She served as precinct committeewoman for the Republican Party. In 1976, she was given the Distinguished Service Award by the Cosmopolitan Club of Aurora. She served as president of the Illinois Library Association, receiving the group's Honorary Librarian citation in 1963. Plain supervised the construction of the present Aurora Public Library. Survivors included one daughter, Jeanne Plain Goss of Arcadia, California. "In her death, we celebrate her life. It was well lived, spent bringing together ideas and people," said the Reverend David Diercks at her funeral.

Herschel and Eva Luckinbill

"Until you act on your decisions, they are only wishes," suggests the poet. With adroit leadership, Eva Davenport Luckinbill and the library board of directors purchased land and formulated plans to build a new, state-of-the-art library in downtown Aurora. Eva was born on July 4, 1947, on the north side of Chicago. She graduated from Roosevelt High School and completed a bachelor of arts degree in education at Northeastern Illinois University, then earning her master's degree in library science from Dominican University. She wed Herschel on June 17, 1967. They are the parents of Glenn, Lorrie, and Brad. They have seven grandchildren. Eva became head librarian of the Aurora Public Library in April 1999. Herschel was born in Oklahoma City on August 2, 1945. He graduated from Plainview High School and joined the Navy on February 23, 1964. He served on the USS *O'Brien* (DD-725), making two tours of duty to the shores of Vietnam. Herschel was president of the Oswego Jaycees, coached pony softball, has been a member of the American Legion and Optimist Club, and has volunteered with Honor Flight Chicago. Herschel presides over the Fox Valley Veterans Breakfast Club, serves on Aurora's Fourth of July parade committee, and has made 25 trips to Washington, DC, as guardian for World War II Veterans on Honor Flights.

Kay and Steve Hatcher

Kay (b. 1945) attended the University of Illinois and earned a corporate communications certificate from Boston College. She was the community relations director for the Fox Valley Park District from 1988 to 1993 and director of external affairs for Illinois Bell from 1993 to 2003. She is president of Reputation Management, Inc., since 2003 and a Republican member of the Illinois General Assembly, representing the 50th District of Kane, Kendall, and LaSalle Counties. Kay has been president of the Kendall County Forest Preserve and served on the Kendall County Board as well as the Oswego School Board. Her honors of the past 20 years include 2009 Legislator of the Year from the Metro West Council of Government and an award of Excellence in Government from the Illinois State Crime Commission. She was named a YWCA Woman of Distinction in 2003 and an outstanding Illinois Woman in Government in 2002. Kay has served on more than 30 local boards, including the Rotary (as president) and the Valley Industrial Association Board, where she was the third woman in a century to serve. She and Steve have four children and six grandchildren. Steve (b. 1947) graduated from Southern Illinois University in 1972 with funding from the Veterans Administration, after surviving wounds during the Vietnam Tet Offensive in 1968. He led the chamber of commerce in Springfield, Missouri, and Kearney, Nebraska, and for the city of Aurora. He currently is president and chief executive officer of the Oswego Illinois Chamber. His benchmark efforts have earned him the respect of community and professional peers, who elected him president of Chamber of Commerce Associations in both Nebraska and Illinois. His local community presence has included the River Valley Workforce Investment Board, Communities in Schools, Provena Fox Knoll Board, Boy Scouts of America, and the United Way as campaign cochair with Kay.

Mayor Paul Egan
Paul Egan (1898–1968)—who came off the state's unemployment compensation rolls to win election with a 3,000-vote plurality and serve eight stormy years as Aurora's mayor—died of cancer on August 22, 1968. Mayor Egan was an eccentric who hired a woman wrestler as a bodyguard, discharged the entire 68-man police force and named a 28-year-old, red-headed woman as chief, had fist fights with other public officials, and compared Lenin with Jesus at a Palm Sunday city council meeting. He was a stocky, talkative former reader of electric meters. He won election to a four-year term in 1953, while he was receiving state relief of $27.50 a week; he began earning $8,000 a year as mayor of a city with a population of 57,000. Thus, he was able to support his wife, Pat, and five children. He had been born in Ottawa. He sold newspaper advertising space for 23 years and later set up his own newspaper, the *Aurora Economist*. Pat said he was "the catch of the town . . . good-looking, witty, and of course, strong-willed." They had married in 1939 at a grand affair at the Leland Hotel in Aurora. "I would tell him 'Oh, Paul, don't do that' and he always would." Mayor Egan's body was donated to medical science. This photograph shows presidential candidate John F. Kennedy in 1960 with Mayor Egan in Aurora receiving the key to the city. (Courtesy of Aurora Historical Society archives.)

State Representative Linda Chapa LaVia

LaVia was born on August 16, 1966. She graduated from East Aurora High School and earned her bachelor of science degree from the University of Illinois at Chicago. Linda graduated ROTC and is an Army veteran. She is a licensed real estate broker with Chapa Realty and has been active in the Greater Aurora Chamber of Commerce. Linda's 83rd State of Illinois district includes most of the city of Aurora. She was elected to the Illinois General Assembly in 2003, the first Hispanic Democrat to win a seat outside of Cook County. Linda married Vernon LaVia (b. 1963) on June 25, 1992, and they have two daughters, Veronica and Jacqueline.

Mary Lou and Benito O. Chapa Sr.

As the first Hispanic woman realtor in Aurora, Mary Lou Chapa was a trailblazer. She has been a member of the Tri-County Association of Realtors, the State Board of Realtors, and the Hispanic and Greater Aurora Chambers of Commerce. She was born on February 25, 1938, to Jesus Olivarez and Antonia Barrientes in Robstown, Texas. She married Benito on July 31, 1957. They have four children: Fernando, Esmeralda, Benito Jr., and Linda. Mary Lou was named a YWCA Woman of Distinction in 2005 and grand marshal of the Fourth of July parade in 2009.

Aurora Township Youth Commission

It was resolved on June 7, 1984, that the Aurora Township supervisor shall appoint members of the commission. Shown here are, from left to right (first row) Paul Patricoski (vice chair), Jo Fredell Higgins (chair), Christina A. Campos (Aurora Township supervisor), Justine Kopytov (director of the youth center), and Antonio Gasca (District No. 131); (second row) Delores Hicks (township trustee), Diane Christian (Aurora Public Library), Vicki Marquez (youth center), Scot Thurman (Wayside Cross), and Belinda Brooks (Quad County Urban League). (Courtesy of Dave Chrestenson.)

Sen. Chris Lauzen

Chris Lauzen was born to Leo and Violet Moldovan Lauzen on December 30, 1952. He earned his bachelor of arts with honors from Duke University in 1974, and graduated from Harvard with a master of business administration degree in 1978. Lauzen earned his certified public accountant (CPA) degree from the University of Illinois. He wed Sarah Longley on July 25, 1981, and they have four sons, Ted, Elliot, Hans, and Robbie. He served as a Republican senator in the Illinois State Senate from 1992 to 2012. In 2012, Lauzen won the Republican primary for Kane County Board chairman with 70 percent of the vote, taking 29,818 votes.

William "Bill" Catching

William Catching was born on May 29, 1968, to William and Therese Catching. He graduated from Bishop Eustace Preparatory High School in Pennsauken, New Jersey. After graduating from the Medill School of Journalism at Northwestern University in 1990, Catching moved to Aurora to work at the *Beacon News*. He has served as a board member of Communities in Schools, West Aurora Volunteers for Education, and Aurora Cease Fire. He currently serves as a trustee for Aurora Township. Catching is president of Catching Communications. He has three children: William, Zoe, and Caleb. (Courtesy of Donnell Collins photography.)

Mayor David L. Pierce

Pierce was born in 1947 in Aurora. He wed Susan Traman in 1975, and they have one son, Ryan. Pierce graduated from Marmion Military Academy in 1965 and St. Procopius College, now named Benedictine University, in 1969 with a bachelor of arts in political science. He studied at the University of Illinois Law School from 1969 to 1971. He served as Kane County clerk from 1974 to 1985. Pierce was elected mayor of Aurora in 1985 and served until 1997. He was appointed as a special assistant to the deputy director at the Illinois Department of Commerce and Community Affairs, serving from 1997 to 1999, and served as village administrator for the Village of South Barrington from 1999 to 2010. Mayor Pierce received honorary doctorate degrees from Benedictine University and Aurora University.

Mayor David L. Stover
Born on March 28, 1947, Stover graduated from West Aurora High School in 1965. After military service, he took a job with the FBI in its Chicago office. In 1969, he began his career with the Aurora Police Department, and was named Patrolman of the Year in 1971. After serving as Aurora's chief of police, he was elected mayor in 1997 and held that post until 2005.

Arlene Helen Fichtel Shoemaker
Amid the swamp white oak and shagbark hickory trees of the Arlene Shoemaker Forest Preserve, visitors can enjoy nature's beauty. Shoemaker was born on June 4, 1931, to Christian Fichel and Leona Clara Schratz Fichtel. His parents, John and Katharina Fichtel, moved to Aurora in 1858. Arlene wed Jerome Shoemaker (1930–2005) on February 5, 1949, and they raised 10 children. There are now 21 grandchildren. Arlene served on the Kane County Board from 1980 to 1992 and was president of the Kane County Forest Preserve Board.

Marilyn Michelini
Michelini was born on January 4, 1931, to Ray J. and May Boozel. She served on the committee that led to the formation of Waubonsee Community College. Marilyn's husband, Richard (1931–2003), was regional sales manager for Ciba-Geigy for 33 years. He served on the Montgomery Village board and was chair of the Aurora Township Democratic Party. Their children are Tom, Marie, Mark, Lisa, and Christine. Michelini received her associate degree from Waubonsee and worked for a decade at the YWCA in Aurora as outreach director. She served on the boards of the Fox Valley United Way and the Aurora YWCA. She has been a volunteer counselor for domestic violence through Mutual Ground. Michelini has been a Kane County Board member and has been Montgomery Village president since 2001. (Author's collection.)

Jim Oberweis

The Oberweis dairy was founded in 1927 by Jim Oberweis's paternal grandfather, Peter J. Oberweis. Jim (b. 1946) graduated from Marmion Military Academy, received his bachelor of arts from the University of Illinois, and his master of business administration degree from the University of Chicago. In 1986, he bought Oberweis Dairy from his Aunt Marie and his older brother John. Aunt Elaine Oberweis had taken charge in the 1980s and rebuilt the dairy as a high-end brand. The 2008 revenue was about $72 million. Oberweis Dairy went from 50 employees to over 1,000 employees today. Oberweis Securities was bought out, and Jim Oberweis then repurchased a small mutual fund he had begun. Today, Oberweis Asset Management manages somewhere between $1.5 and $2 billion. Oberweis has served on the Republican Party's state central committee, representing the 14th Congressional District. Jim won the 25th District State Senate seat Republican nomination in March 2012. Jim is the father of five and the grandfather of 11.

Marie Oberweis

Peter J. Oberweis found that he had too much milk, so he began selling it to neighbors in 1915. A quart of milk in 1939, home-delivered, cost 15¢. Skim milk was sold back to farmers for hog feed for 2¢ a gallon. In 1939, Marie (1921–2007), far right, joined the family business. In George Washington's ledgers, there is a notation on May 17, 1784, for the expense of a "crème machine for ice." (Author's collection.)

OBERWEIS DAIRY
--- "O for the Sweetness of It"

SCITECH
--- "A Science Museum"

WHITE EAGLE CLUB

AND MORE...

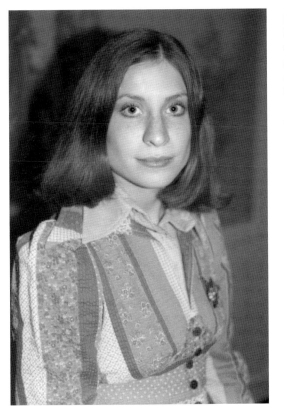

Karen Fullett-Christensen
Fullett-Christensen (b. 1951) was born in Chicago to Dorothy and Ben Fullett and attended Mather High School. She is a 1972 magna cum laude graduate of NIU, with a bachelor of arts degree in social sciences. Fullett-Christensen has lived in Aurora since 1999 with her husband, Larry. She has two adult daughters, Miriam and Kira. Fullett-Christensen is currently the manager of the City of Aurora's neighborhood redevelopment division. She is a member of Temple B'nai Israel.

Gina Heinz Moga
Moga was born in Aurora on June 10, 1959, to Peter and Sara Heinz. She graduated from East Aurora High School in 1976. In 2007, Moga was appointed manager of the Mayor's Office of Special Events. Moga serves on the board of the Illinois Special Event Network and on the Aurora Area Convention and Visitors Bureau. In 2011, Moga became the development coordinator for the city of Aurora. She wed Zack Moga on November 19, 1977, and they have three grown children, Daniel, Alexandra, and Andrea, and one granddaughter, Ava. (Courtesy of Donnell Collins.)

Harry Burt Stoner

When Harry Stoner (1906–1976) told his mother in 1946 that he'd bought a farm on which he was going to develop a shopping center, she asked, "What's a shopping center?" The development, along North Lake Street, became known as Northgate Shopping Center, and had 52 stores occupying 500,000 square feet. Stoner was known as the "Father of Lake Street," because for years he promoted the area as the "Golden Mile" of Aurora. Stoner was the chairman of the board of Valley National Bank and a director of the Aurora Foundation. He was president of the Aurora Library Board from 1958 until 1971 and was a member of the advisory board of Mercy Center Hospital. He was president and founder of the Stoner Manufacturing Company, known for its vending machines. The company manufactured 20-milimeter cartridge cases during World War II. Death came on March 29, 1976. His wife, Ann, son, David W. Stoner (1946–2001), and daughter-in-law, Gyda Otten Stoner (b. 1942) survived him. David W. and Gyda had two sons, William and David, and five grandchildren.

Melvin William "Bud" Meyer

"Bud" Meyer was born on February 1, 1923, to William F. and Lena Einsiedel Meyer of Aurora. He was a lifetime member of St. Paul's Lutheran Church, where he served as chairman of the congregation and on the building committee for the Orchard campus. After service in the US Army Coast Defense Command during World War II, he returned to Aurora and married Wanda Hammond Meyer. They had two sons, William and John. Meyer cofounded the plumbing supply firm William F. Meyer Company, for which he served as president until 1987. He served on the Aurora Planning Commission for 12 years. He was twice president of the Aurora Chamber of Commerce, past president of the Aurora Little League, and national president of the American Supply Association. He was the recipient of the Aurora Chamber of Commerce's highest award, the Albert M. Hirsch Award, and was awarded an honorary doctorate of humane letters from Aurora University. Bud was grand marshal for the 1995 Fourth of July parade. After the death of his wife, Wanda, in 1999, he married Jean Roesch Selfrid on January 1, 2001. He was chairman of the Aurora Civic Center Authority and Paramount Arts Centre as well as chairman of the Provena Mercy Hospital Foundation. For his service on North Island Centre Authority Board, the ballroom was named in his honor. Meyer died on July 6, 2003.

Dennis P. Wiggins
Denny Wiggins was born on December 28, 1940, the youngest of 11 children born to Donald and Rachel Cox Wiggins. He graduated from East Aurora High School in 1959 and served in the US Army Reserve from 1963 to 1969. He is currently executive director of Joseph Corporation of Illinois, Inc., and served as Aurora Township supervisor from 1973 to 1997. While supervisor, he established the senior services committee, implemented a Dial-A-Ride program for seniors and handicapped persons, opened the Aurora Township Senior Center, and established the Township Youth Commission and the Township Youth program. He established the Summer Free Lunch Program for Youth in 1992 and assisted in establishing adult day care, Carrier Alert, Skills Bank, the Neighbor to Neighbor program, and the Senior Shared Housing. Wiggins has been chair of the Aurora Township Republican Central Committee and presidential advisor and board president of the Kiwanis Club. He was named Humanitarian of the Year with the Lyle E. Oncken Community Service Award in 1995 and received the Distinguished Agency Award from the Northeastern Illinois Area Agency on Aging. Wiggins received the President's Award for Leadership from the Township Officials of Illinois. He has been involved with the Kiwanis Club for 26 years, the Aurora Sports Boosters Club, the American Cancer Society, and the DuKane Valley Council. His wife is Carol Weiland, and his children are Christopher, David, Kimberly, Kelly, and Erica. (Courtesy of Greg Stangl.)

CHAPTER FIVE

Gracious Golden Apples

The hand that gives, gathers.

—Psalm 41

Dr. Christine J. Sobek
Dr. Sobek earned her bachelor's degree with highest distinction from Purdue University, her master's from Michigan State University, and her doctoral degree from NIU, DeKalb. She is a member of Phi Beta Kappa and an honorary member of Phi Theta Kappa International Honor Society. Dr. Sobek has more than 30 years of community college administrative experience. She became the fourth president of Waubonsee Community College on July 1, 2001. The college has a network of four permanent campuses located across the 600-square-mile college district and extensive online offerings. Among her awards are the Outstanding Alumni Award from NIU, the Shirley B. Gordon Award of Distinction by Phi Theta Kappa International Honor Society, and a Quad County Urban League Lifetime Achievement Award. She currently serves as a member of the Rush-Copley Medical Center Board of Directors, the Valley Industrial Association Board of Directors, the Conservation Foundation Board of Trustees, and the Marmion Academy Board of Lay Trustees. Other leadership roles include service on the American Council on Education Commission on Racial and Ethnic Equity, the American Association of Community Colleges Sustainability Taskforce, and the AACC Commission on Marketing and Communications. Dr. Sobek and her husband, Dr. Paul Anderson, are the parents of Elliot, Eric, and Amelia. (Courtesy of Waubonsee Archives.)

Dr. William Marzano
Dr. Marzano was born on August 29, 1949, in Chicago to Anthony and Angela Marzano. He attended Our Lady of Charity and J. Sterling Morton East High School in Cicero. He graduated from Morton Junior College, where he was valedictorian, in 1969. He earned his bachelor's degree from NIU in DeKalb and his master's degree from the University of Illinois in educational psychology. He also served as full-time psychology instructor at Illinois Valley Community College. During that time, he completed a doctor of education degree in curriculum and instruction from Illinois State University. On June 15, 1974, he married Mary Hudetz of Warrenville. They have three children, Anthony, Gwendolyn, and William Jr. Dr. Marzano worked for the next 15 years as a human resources manager and in 2000 began as an administrator at Waubonsee Community College. He has served as dean for communications, humanities, and fine arts, as the assistant vice president of community development, and in 2011 became dean for social science and education. Dr. Marzano is a member of the Illinois Council of Community College Administrators, past board chair of the Aurora Regional Chamber of Commerce, board member of the Aurora Economic Development Corporation, and past president of the Exchange Club of Aurora. He is the recipient of the McMillan Award for Extraordinary Commitment to Training, Association for Graphic Arts Training and Graphic Arts Trainer of the Year Award in 1997. (Courtesy of Waubonsee Archives.)

Dr. Rebecca L. Sherrick
Dr. Sherrick, was born in 1953 in Carthage, Illinois, and earned a baccalaureate degree in history, graduating summa cum laude, from Illinois Wesleyan University. She was awarded an honorary doctorate by her alma mater in 2001. She completed a doctorate degree in history from Northwestern University in 1980. Dr. Sherrick has served since July 2000 as the 13th president of Aurora University. Aurora College began in Mendota in 1893 and moved to the new Aurora campus on April 3, 1912. (Courtesy of AU archives.)

Dale Richard Von Ohlen
Dale Von Ohlen was born on June 4, 1921, to Everett and Lillian Dittman Von Ohlen. He graduated from East Aurora High School in 1939. Dale and partner George Goeltz operated Hansen Mattress Co., where Von Ohlen worked until his 1990 retirement. He wed Marie Lloyd in 1951, and they had four children: Todd, Scott, Ellen, and Neal. He was on the steering committee that became Waubonsee Community College, and he served as president of its first board of directors. Recognizing his years of leadership, Waubonsee named the music and art building Von Ohlen Hall. Dale enlisted in the US Army in 1942.

Dr. Sherry Rosalyn Eagle
Dr. Eagle was born in Sioux City, Iowa, on June 17, 1948, to Irving and Rose Mittleman. She wanted to be an opera star. She wed Howard Charles Eagle (1946–2006) on September 8, 1967, in Sioux City. They had one daughter, Roseanne Michelle Eagle, who was born on August 29, 1977. Dr. Eagle received her bachelor of arts with honors from the University of Illinois, and earned her doctor of education degree in leadership and educational policy studies with honors in 1994 from NIU, DeKalb. Dr. Eagle is currently executive director of the Institute for Collaboration at Aurora University. She was superintendent of schools for West Aurora District No. 129 from 1993 to 2005. Under her leadership, the district designed and constructed two new elementary schools, two new middle schools, five school additions, and renovations of all district buildings, acquiring land and equipment totaling over $150 million. Among her many awards are the Illinois State Board of Education Break the Mold Award in 1995, the University of Illinois Alumni Achievement Award in 2006, the Aurora Public Library Foundation Light of Learning Award in 2005, an Honorary Doctor of Humane Letters from Aurora University in 2004, the Quad County Urban League Leadership award in 2002, and a YWCA Woman of Distinction Award in 1993. Sherry has been a board member of the Quad County African American Chamber of Commerce, the YWCA, and has served on the Governor's Task Force for Education.

Gerald Lubshina

Jerry Lubshina (b. 1941) was born in Ottawa to Eveline and Joseph Lubshina, whose grandparents had emigrated from Croatia. They moved to Aurora in 1963. He received his degree in history from NIU, DeKalb. He then began a 38-year teaching career with East Aurora District No. 131. He married Merrilee Meling, and they had four children: Stephanie, Melanie, David, and Marie. They have two grandchildren, Carys and Joshua. Lubshina was selected the 1990 Teacher of the Year at East Aurora High School. He has taught religious education classes for 21 years at St. Rita of Cascia Church and has served the city of Aurora since 1980 on the block grant committee.

Mabel C. O'Donnell

Mabel O'Donnell was born in Aurora on March 21, 1890, and graduated from East Aurora High School and the University of Chicago. She taught primary students, was a primary grade supervisor, and was a curriculum coordinator at East Aurora District No. 131 before becoming the head of the reading department of Row, Peterson and Company in 1946. While there, O'Donnell was the author of the Alice and Jerry books. In 1965, the new school erected on Reckinger Road was named for her. She died on December 14, 1985. (Courtesy of Dave Chrestenson.)

Grace Nicholson (LEFT)

Edward and Maria Nicholson welcomed their daughter Grace in 1879. She graduated from West Aurora High School and Northern Illinois State Normal School in DeKalb. In 1902, she began her teaching career in Montgomery, for the last 15 years of which she served as principal of Montgomery School. In 47 years as an educator, she missed one half-day of school for a broken wrist. In 1962, the school was renamed the Grace M. Nicholson School. Florence Ward wrote, "Beautiful in form and feature, lovely as the day, can there be so fair a creature formed of common clay?" Nicholson died on November 29, 1966. (Courtesy of Veronica Radowicz.)

Dr. James Rydland

Dr. Rydland was born on October 27, 1949, to Simon and Frances Rydland. He graduated from the University of Washington, Seattle. He wed Jo on September 11, 1976, and they raised two children, Kelsey and Morgan. Dr. Rydland has been superintendent of District No. 129 since July 2005. (Courtesy of West Aurora District Archives.)

Dr. Michael Edward Sestak

Born in Thayer, Illinois, to Thomas Paul and Helen Monica Bednar Sestak on August 8, 1930, Michael Sestak was the ninth of 10 children. His parents were Czechoslovakian immigrants, and neither had more than a second-grade education, but they placed a high priority on their children's learning. Dr. Sestak graduated with his master's degree from Western Illinois University in 1953 and earned his doctorate in educational administration from the University of Illinois in 1964. He wed Mary Ann Timko on August 17, 1957, and they have five children and 14 grandchildren. Mike's career spanned five decades as teacher, coach, principal, college professor, and as the superintendent at Mooseheart Child City. Mike was named Citizen of the Year in 1977 by the Aurora Lions Club and received the Pilgrim Degree of Merit, awarded by the Loyal Order of Moose. He served on the Greater Aurora Chamber of Commerce, the Water Advisory Board, the Aurora Civil Service Commission, and the board of directors of Annunciation Church. Mike was a visionary and a problem solver. He was posthumously awarded the diocesan Bishop Arthur J. O'Neill Award. Dr. Sestak passed on June 5, 2002.

Edna M. Rollins
Rollins was born in Aurora on April 30, 1916, to William and Annie Hodgetts Rollins. She graduated in 1933 from East Aurora High School and later began a 43-year career with the East Aurora District No. 131. She was the director of financial and administrative services and the school district treasurer before retiring in 1979. At the time of her retirement, she was one of two women in the country holding the Registered School Business Administrator certificate. When the Edna M. Rollins Elementary School was dedicated in October 1990, it was Edna Rollins Day in Aurora. Rollins died in 2010. (Courtesy of the Community Foundation of the Fox River Valley and Rollins School.)

Richard "Dick" Schindel
Edmund and Rita Schindel of Aurora welcomed their son Richard on December 14, 1947. Schindel graduated from East Aurora High School in 1966 and from the University of Illinois with a bachelor of science degree in marketing in 1970. On June 13, 1970, he married Susan Roadruck. They have two daughters: Laura and Joanne. In 1973, he was hired with East Aurora District No. 131. He received his master's degree in curriculum from National Louis University in 1989. Schindel retired in 2004. He was selected to the East Aurora Hall of Fame in 2008 in the category of Athlete/Coach/Teacher. Schindel began his business, Dick's Mini-Donuts, in 1996. On April 26, 2012, Dick was inducted as an East Aurora High School Distinguished Alumni.

Sharon Stredde
Stredde (b. 1946) has been the president and chief executive officer of the Community Foundation of the Fox River Valley (formerly the Aurora Foundation) since 1985. Under her direction, the assets of the organization have grown from $3.7 million to $60 million. In the early 1980s, Stredde was the development director of the Aurora YWCA's fundraising campaign to build a new facility. She also served as a member of the board of education for West Aurora School District No. 129 from 1985 until 1989. Sharon was born in Aurora and is a graduate of West Aurora High School. She and her husband, Edward Stredde, are graduates of the University of Illinois. Edward retired in 2004 as the vice president of international technical support for Lucent Technologies. The couple has one child, Robert, who is the owner and president of Technical Theatre Services, Inc. (Courtesy of the Stredde family.)

GAR Memorial Hall

In 1885, it became necessary for the library to build a new home and to give the GAR Post 20 a permanent assembly hall. The addition was erected at a cost of $5,800. This building is included in the National Register of Historic Places. A $2-million restoration project began in 2008, and the building is expected to be completely restored and opened by 2014. Shown in this GAR Commission photograph are, from left to right, (first row) Dan Barrerio (director of city's community development), Perry Slade, and Jo Fredell Higgins (chair); (second row) Rena Church (director/curator of the Aurora Public Art Commission and the GAR Hall Museum), Jay Harriman, Ron Stone, Bill Noltemeyer, and John Heinz. Missing from the photograph is Michael Sawdey. (Photograph by Dave Chrestenson.)

CHAPTER SIX

Ordinary People, Extraordinary Efforts

The morning hour has gold in its mouth.

—Dutch proverb

Daniel D. Dolan Sr., Dolan & Murphy

As Aurora celebrates its 175th anniversary, it notes Daniel Douglas Dolan Sr. (b. 1931), architect and visionary of Dolan & Murphy, Inc. Real Estate. Dolan graduated from East Aurora High School in 1949, then joined the Navy, serving on the USS *Knapp* (DD-653) during the Korean conflict. Returning home, he served for five years as an Aurora fireman. After selling real estate with the Liberty Realty Firm, which dealt with commercial investment, he left, and in 1965, formed Dolan & Murphy with partner James O. Murphy. Their first large project developed 365 acres around the interchange of I-88 and Farnsworth Avenue for business and industrial development. A joint development by McDonald's and Del Ray Farms Grocery resulted in a dozen parcels in the heart of a residential area being combined. This increased the tax base on the property manyfold while creating 150 full-time and part-time jobs. Dolan & Murphy, Inc. has become a leader in property management, with one million square feet of office and retail space, more than one-half million square feet of industrial space, and more than 40 active land partnerships in its portfolio. Dolan has garnered numerous awards, including the Distinguished Alumni Award, class of 1949, East Aurora High School, in 2011, and the Bishop Arthur J. O'Neill Award for his efforts in building the new Aurora Central Catholic High School. Dolan was named grand marshal for the City of Aurora's 2004 Independence Day parade. He has sponsored Little League teams for the past 40 years and sponsored the Dolan & Murphy Shamrock Fast Pitch Softball Team. He has established the Daniel D. Dolan Family Advisory Endowment Fund with the Community Foundation of the Fox River Valley as a permanent resource for his family's philanthropic interests. (Donnell Collins photography.)

The Dolan Sons and Grandchildren
Shown in this photograph are, from left to right, (first row) Daniel T. Dolan, Daniel D. Dolan Sr., and Brian K. Dolan; (second row) grandson Daniel John Dolan, Ryan J. Dolan, and granddaughter Wendy Brennan. Brian has been engaged in the marketing of industrial and commercial real estate since 1977. In February 1988, he received his CCIM designation. Brian is the chair of the board of directors for the Marie Wilkinson Food Pantry, past president of the Aurora Central Catholic Board, and a member of the Aurora Kiwanis. He is also chair of the Aurora Economic Development Commission. Daniel T. has been with the family firm since 1973. In 2011, he received the Community Champion Award by the Compassion Foundation, and was the 2011 recipient of the Outstanding Corporate Citizen Award for extraordinary support of the Marie Wilkinson Pantry. Daniel has been cochair and chair of the advisory panel of CASA. He was cochair of its endowment campaign and has served on the board of directors and was past chair for the Aurora Economic Development Commission. Ryan graduated from Aurora Central Catholic High School in 1991 and received his real estate license in April 1996. He has been involved with the 400-plus-acre farm sale in Bristol Township to Moser Enterprises of Naperville, Illinois. (Courtesy of Donnell Collins Photography.)

Ken Nagel

Nagel was born in Aurora on August 30, 1942, to Erwin and Olive Stark Nagel. Their last name means "nail" in German. He graduated from St. Nicholas School and Marmion Military Academy, and then joined his father's firm, which became known as Fox River Foods. Nagel's partner at Fox River Foods is Frank Karabetsos. Today, Fox River Foods has 500 employees and 5,000 customers, with annual sales of $400 million. Every day, 100 trucks deliver food loads in seven states, which translates to 60,000–70,000 cases of food product delivered daily. Nagel married Carole Kotar in 1967, and they have four children, Brenda, Molly, Jason, and Betsy. The couple has 12 grandchildren. Nagel's leadership garnered him the Spark Plug Award from the Chicago Food Service Marketing Club in 1983, and he was chosen for their hall of fame in 1990. Nagel was named Person of the Year by the Illinois Foodservice Association in 2001. In 2008, the company received the International Foodservice Manufacturers Association top honor, the Excellence in Distribution Award. Nagel has served on the Marmion Board of Trustees for the past 20 years and received the Centurion Award from Marmion in 2004.

Frank H. McWethy
McWethy graduated from West Aurora High School in 1901 and completed his college education at the University of Wisconsin. He had been born in Aurora on April 25, 1882, of Scotch ancestry. McWethy wed Gertrude Erickson in 1906, and they had three children: Margaret, Frank, and James. Gertrude died on January 23, 1940. After Frank's sudden death on December 21, 1944, Stephens-Adamson Company was credited with designing a "people-mover" that led to the mobile sidewalks later used by airports and subways around the world. Of the six generations of McWethys living in Aurora, from his eight grandchildren, McWethy is saluted for his contributions to manufacturing and for bringing lasting international acclaim for Aurora. (Courtesy of Nancy Smith Hopp.)

Caterpillar Tractor Company
The Caterpillar plant opened in 1958, when Aurora's population was 50,600. Bobette Keasier was the first local resident hired. Shown here is the first 955 Traxcavator sold to Geneva Construction Company on October 19, 1958, from the Aurora/Montgomery plant. A.N. Whitlock (left), the first plant manager, is shown with an unidentified representative from Geneva Construction Company. (Courtesy of Dave Chrestenson.)

Schaefer's Greenhouse

Frank C. Schaefer (1879–1952) emigrated from Berlin, Germany, almost penniless in 1903. He wed Ida Haase in 1905, and they had eight children and 28 grandchildren. He did greenhouse work in Aurora for Judge John Newhall, who loaned him money to start the Aurora Greenhouse Company with a partner in 1907. Frank became a citizen in 1910. His business acumen laid the groundwork for a family business that now spans over 100 years. Schaefer's Greenhouse is the premier florist in the Fox River Valley.

Prisco's Fine Foods

"Industry is fortune's right hand," and this is applicable to grandfather Anthony Prisco Sr., shown here. Initially, Prisco purchased a pickup truck and drove door-to-door in the early 1930s selling groceries. He and his wife, Mary, also sold goods from the front room of their home on Bishop Avenue. Today, Robert, the current president and the oldest of Tony Jr.'s eight children, and his wife, Georgette Prisco, and four other Prisco family members continue the tradition of excellent foods and quality service at Prisco's Fine Foods. They include Tony G., Margaret, Beth Prisco Guzauskas, and Anne Louise Prisco Strong. (Courtesy of Martha Prisco.)

Leonard Douglas
Leonard Douglas was born in Aurora at Copley Hospital to Irene and C.A. Douglas on December 19, 1933. He attended Bardwell School and graduated in 1952 from East Aurora High School. He took classes at Aurora College for a term and then joined the Navy, serving from 1953 to 1956. On June 30, 1956, he married Marilyn Barter at the First Methodist Church in LaMoille, Illinois, near Mendota. Douglas worked for the post office for the next 11 years, and the couple had three children, Mike, Jim, and Christine. There are now nine grandchildren. Douglas opened Douglas Floor Covering in their home and worked from there between 1963 and 1969. A building was erected on Route 25 in North Aurora, and in 1982 son Mike took over the business. In 1987, they opened the Lincolnway location until 1995, when they moved to the Randall Road location. The business is now on Orchard Road, and the warehouse is on Alder Drive. Both sons have been involved in the business since 1980. Leonard served on the Kane County Board from 1986 to 1990 and has been a North Aurora Lion's Club member for the past 50 years.

CHAPTER SEVEN

Excel in the Grace of Giving

They shall not grow old, as we that are left grow old. Age shall not weary them, nor the years condemn.
At the going down of the sun and in the morning, we will remember them.

—Laurence Binyon

Fr. David Engbarth
Joseph and Frances Engbarth welcomed their baby son on August 24, 1948, in Rockford, Illinois. He attended St. Anthony's grade school and Boylan Central High School, graduating in 1967. Father Engbarth received his bachelor of arts degree in sociology from Loyola University and studied for four years at St. Mary's Seminary in Baltimore, Maryland. He was ordained on November 20, 1976, in Freeport, Illinois. Father Engbarth has served at St. Mary's in Sterling, Holy Angels in Aurora, St. Joseph's in Elgin, and came to St. Nicholas in 1993 to 2005. After a year of sabbatical, he returned to Aurora and became pastor of Our Lady of Good Counsel in 2006. He began, with Pastor Dan Haas, the Prayer Coalition for Reconciliation in 1994 to raise voices for peace in Aurora, which continues to this day. More than 250 prayer vigils have been held. The Good Counsel parish began on April 26, 1909, with Fr. Leon M. Linden as the first pastor. Sermons in German were discontinued on June 11, 1916. The school opened on September 10, 1910, staffed by the school sisters of St. Francis. There were eight graduates in the first graduating class.

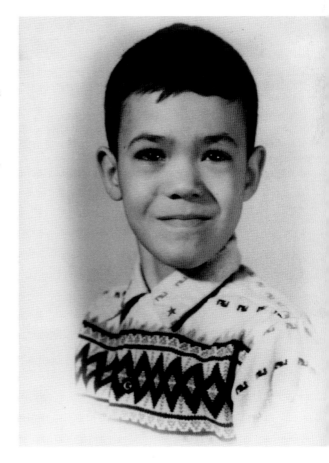

Sr. Dorothy Burns, RSM
Dorothy Burns was born in Chicago on October 28, 1933, to Catherine and James Burns. Her parents had emigrated from Ireland in the late 1920s. Burns had one brother, James Michael, and one sister, Patricia Ann. She graduated high school at St. Patrick's in Des Plaines and attended DePaul and St. Louis Universities as well as Notre Dame, studying financial management for hospital administrators. She professed her vows as a Sister of Mercy of the Americas at Queen of Martyrs Church in Chicago on August 16, 1960. Sister Dorothy began working at the old St. Joseph's Hospital in 1967 in admitting and on the switchboard. She served as vice president and later chief executive officer of Mercy Hospital from 1990 to 1996. She then was administrator of the Fox Knoll religious community on Lake Street from 1997 to 2001. Sister Burns now resides at McAuley Convent in Aurora.

1979, Tony Zalanas and Annie Campbell receive their 25-year service pins.

Sr. Rita Meagher, RSM

Rita Meagher (1923–2011) was born in Chicago to Walter and Loyola Dougherty Meagher. She attended Mercy High School and entered the Sisters of Mercy in 1944. Sister Rita received her bachelor of science degree in nursing from St. Xavier University in 1953. She earned a master's degree in hospital administration from St. Louis University in 1959 and an honorary doctorate from Aurora University in 1983. She was the chief executive officer at St. Joseph Mercy Hospital and oversaw the building of the new general hospital. She helped with the merger of Mercyville Institute of Mental Health and Mercy Medical Center into Mercy Center for Health Care Services, where she served as chief executive officer until 1990. In 1991, she became the planned giving officer for Mercy Center and then served as a volunteer until her retirement in 2004. She was the first woman chair of the Illinois Hospital Association and a recipient of the YWCA Woman of Distinction Award. The Sister Rita Heart Center, named in her honor, treats cardiac patients in Aurora. Sister Rita was survived by her sister, Loyola Kane, and nieces and nephews. Shown from left to right are Ann Campbell, Sister Rita, and Dr. James Habbager.

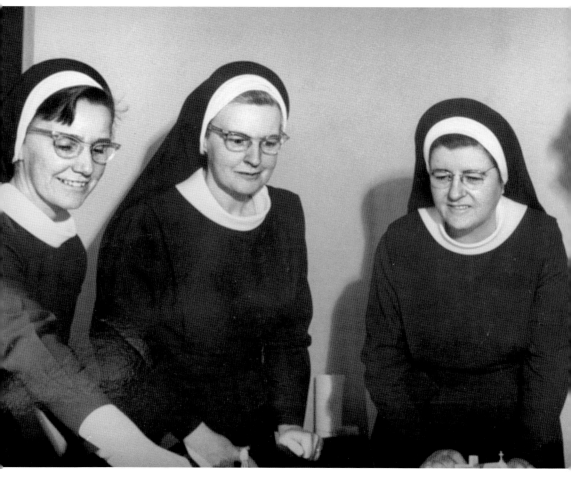

Sr. Mary Assumpta Buckley, RSM

Margaret Buckley was born on July 2, 1915, in Chicago, Illinois, to Francis and Laura Agnes Flavin Buckley. When Buckley was 15 years old, her father died suddenly, leaving her mother to support and care for the family. Buckley attended Madonna High School and lived with the sisters. When she was old enough, she became Sister Margaret and later took the name Sr. Mary Assumpta when she professed her vows in 1939. She graduated with a bachelor of science degree in nursing from St. Xavier College in 1957. She received her master's degree in hospital administration from St. Louis University in 1962. Sister Assumpta served as hospital administrator at St. Mary Hospital in DeKalb, St. Joseph Mercy Hospital in Aurora, and Mercyville. She was councilor on the Provencal Council of the Sisters of Mercy, Chicago Provence, and was a board member of the Mercy Hospitals in Chicago, Aurora, Marshalltown, Iowa City, Davenport, and Janesville. She served as a member of the board of trustees of the Illinois Hospital Association and the National Association of Private Psychiatric Hospitals. Sister Assumpta was survived by her sister, Mary Mercedes, and nieces, nephews, grandnieces, and grandnephews upon her death on February 5, 2011. Shown from left to right are Sister Petrina, Sister Assumpta, and Mother Paulita.

St. John's United Church of Christ

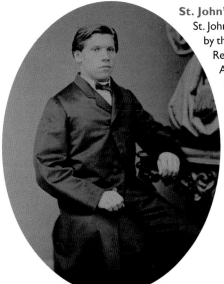

St. John's Evangelical Church was organized on September 5, 1886, by the Reverend Karl Krumm for the area's 12 German families. Rev. Krumm was given $600 a year as a full-time salary. Christ Armbruster was given the contract to build a church on the vacant lot next to the former parsonage, and by September 29, 1887, the building was dedicated. Reverend Krumm is pictured. (Courtesy of Pastor Cyndi Gavin.)

Community Clothes Closet at St. John's United Church of Christ

In the autumn of 2011, the Reverend Cyndi Gavin, pastor, founded a clothes closet with an $18,000 grant from a consortium of private donors. Another of Reverend Gavin's endeavors is the "Mesa de Maria," Mary's Table cooking classes and table spirituality begun in 2009. Hispanic women gather to prepare the meal and share their stories and art in graceful fellowship. Their pictorial history reads, "Blest be God, Who is our bread! Let all the world be clothed and fed!"

Bishop William Haven Bonner

William Bonner was born on September 13, 1921, in Blytheville, Arkansas, to Eddie and Mattie Bonner. He graduated from DuSable High School and Moody Bible Institute. Bonner was awarded an honorary doctorate from the Trinity Hall Theological Seminary. He wed Dorothy Mae Green on May 2, 1942, and they had four children: Enyess, Willie, Charles, and Dorothy. In 1944, Bishop Bonner began church services in Aurora. He and his small congregation broke ground for the Mount Olive Church of God in Christ in 1948. He has received numerous awards, including the 2011 Nia Award for his community leadership from the African American Heritage Board. Bonner has served as pastor for 68 years, enriching his life and many others as well. Proverbs 24:3 reads, "By wisdom a house is built. By understanding, it is established." (Courtesy of Willie Etta Wright.)

CHAPTER EIGHT

Everlasting Evergreen

*Each, for a little moment filling up some little place and thus
we disappear in quick succession, and it shall be so 'till time, in one vast
Perpetuity, be swallowed up.*

—Author unknown

Jeff Scull
In 1984, Scull (b. 1950) began working for Aurora
Township and subsequently established a widely
recognized and award-winning program for youth.
The Aurora Township Youth Center has received
the Outstanding Youth Program Award by the
Illinois General Assembly, the Mayor's Award
in 1991, the Governor's Home Town Award in
1991 and 1997, the Youth Program of the Year
Award from the Association of Illinois Township
Committee, the "Caring for the Future" Award in
1995 from Family Focus, and the Fox Valley Region
Jaycees Excellence in Performance Award in 1990.
Scull was chosen Image Maker by the Aurora
Chamber of Commerce in 1998. He received
his master's at teaching degree in 2001 from
Aurora University. Scull has been president of
the Jaycees, has served on the board of directors
for Breaking Free/Family Support Agency, the
Advisory Board for Family Focus, and as president
of the Illinois Coalition for Community Service
(ICCS). He and his wife, Carla, have been married
since September 4, 1992, and have three children.
Carla teaches at Greenman School, West Aurora
District No. 129, and Jeff retired in 2012 after
30 years as a youth advocate/youth director and
teacher for District No. 131 in Aurora.

John Frederick Schneider

Schneider (1828–1899) came to North Aurora with his family at age six. In 1844, a log cabin was built for a school on the Schneider property. John and his father, John Peter, built the mill in North Aurora in 1862. They were one of the first German families in what was to become Kane County. Eva Schneider (1869–1941) bore Evelyn Winter Tavenner (1906–1983), and Evelyn's daughter Ellen was to move to a farmhouse on Sullivan Road in Aurora in 1946. Ellen went to West High School, graduating in 1954. (Courtesy of Ellen Shazer.)

Ellen Tavenner Shazer

After graduating from West High, Ellen Tavenner worked for Dr. Charles R. Deindorfer, dentist, and then with Phil Schalz, before joining the office of Dr. Walter Sperry, dentist. She met Bill Shazer (1935–2010) on a blind date, and they married on May 7, 1955. Bill worked at Barber-Greene and then Caterpillar for 48 years and retired in 1995. He was a stock run production manager and plant superintendent at Barber-Greene. They had one son, Billy, who was born on September 24, 1958.

Mavis Radi Bates

Mavis Bates was born in Waukegan, Illinois, on June 25, 1948, to Claire Acker and Col. Henry Radi. She received her master of science degree from the Illinois Institute of Technology, and became technical manager in the Bell Labs field representative organization. Her next career move was earning a master of science degree as a licensed acupuncturist in traditional Oriental medicine. Bates has been in private practice, Inner Harmony Acupuncture and Oriental Medicine, since 2005. Bates is the founder and chair of Aurora Green Lights, which organizes the Green Fest events, and is president of the Aurora Kiwanis. She and her husband, John Koranek, have two daughters, Madrid Bates Kaufmann and Deneb Bates.

The Aurora Regional Fire Museum

In 1856, a meeting of volunteer firemen resulted in the establishment of Young America Fire Engine Company No. 1. On January 28, 1882, the fire department became a partly paid organization. In 1894, the city constructed a new Hose House on North Broadway, which is referred to as the former Central Fire Station. The Aurora Regional Fire Museum opened at that location on July 4, 1990. Currently, it is managed by Deborah Davis, executive director, and David Lewis, curator.

Fire Chief Bob Mangers
Bob Mangers was born on August 26, 1925, to John and Lena Mangers. He graduated from Marmion and served in the Navy from 1943 to 1946. Mangers joined the Aurora Fire Department in April 1950, was named chief in 1977, and retired in 1990. He wed Jean Marzuki on April 19, 1952. They had eight children, 10 grandchildren, and three great-grandchildren. Jean passed on March 10, 1992. Mangers was Citizen of the Year for the Noon Lions Club in 1990, the same year that he was chosen Fireman of the Year. He served as a Democratic trustee for Aurora Township from 1997 to 2001. Bob and Jean's daughter Jan Mangers was director of historic preservation for the city of Aurora from 1991 to 2010.

Nancy McCaul
Born in South Holland in 1960, Nancy McCaul graduated with honors and high distinction from Valparaiso University, and earned her master of science degree from DePaul University. She and her husband, Joe, have two sons, Jeffrey and David. McCaul became the leader of the Fox Valley Park District in 2010 following the retirement of the previous executive director, Steve Messerli. Under his tenure, the park district had passed the $44.85-million tax referendum in 2008. The district owns and maintains 2,500 acres of open space, 22 miles of river shoreline, 155 parks, 54 baseball fields, 36 soccer fields, and 35 miles of trails. There are 89 playgrounds, 60 basketball courts, and 29 outdoor tennis courts.

Hilary Kay Brennan

Brennan was born in Cleveland, Ohio, on September 26, 1944. She and Tim Brennan were married on March 26, 1966, and in 1972, they and their two daughters, Jessica and Laura, moved to Aurora. Their son, Robert, was born five years later. They have three grandchildren. Brennan graduated from St. Louis University in 1966 with a bachelor of science degree in psychology and from Aurora University in 1983 with a bachelor of arts degree in accounting. She was YWCA Aurora's director of marketing and development. In 1972, she was chosen a YWCA Woman of Distinction. She serves on the board of Provena Mercy Medical Center Foundation and as governing board chair of Provena Mercy Medical Center, on the Aurora University Board of Trustees, and as board treasurer of the Aurora Area Interfaith Food Pantry. She has also served as president of the Community Foundation of the Fox River Valley, of the United Way, and of YWCA Aurora. In 2011, she was honored with the naming of Provena Medical Center's Hilary K. Brennan Lobby.

Selma Gladney
Marie and John Gladney welcomed their daughter Selma on November 29, 1966. Gladney graduated from East Aurora High School in 1985. She is a 1996 graduate of the Chicago School of Massage Therapy and graduated from Midwest College of Oriental Medicine with a master of science degree in Oriental medicine in 2007. She studied advanced acupuncture in Beijing, China, at the Training Center of China Academy of Chinese Medical Sciences and is a certified advanced master of Tung's orthodox acupuncture. Gladney founded The Emperor's Medicine, LLC.

Paul Patricoski
Patricoski was born on March 24, 1955, at Fort Rucker, Alabama, to Thomas and Marie Patricoski. He married Stephanie Galiardi on August 5, 1978, and they have three children, Adam, Matt, and Amanda. Patricoski received his bachelor of science degree in biology from St. Vincent College and his juris doctorate from the University of Notre Dame Law School in 1981. He is an arbitrator for the 16th Judicial Circuit, Kane County, and is a partner in Dreyer, Foote, Streit, Furgason & Slocum Law Firm. Paul received a Friend of Youth Award in 2007 from Aurora Township and has served 16 years on the youth commission.

Anne Goldsmith

Goldsmith, whose natal date was October 31, 1918, was born to Maurice and Ida Holtzman in Chicago. Anne's love of the theater and arts prompted her to start the Paramount Arts Centre endowment in 1983. The fund grew to more than $2 million in a few years. She helped to organize the original series of gourmet picnics in Aurora. In 2008, she was honored by the Exchange Club with the Hal Beebee Book of Golden Deeds Award. She wed Zalmon Goldsmith on December 24, 1939, and they had two children, Bruce and Ellen, and five grandchildren. Zalmon was born on February 5, 1915, in Aurora to Jacob and Pearl Goldsmith. He was a prominent attorney who served on the boards of local organizations. He passed on March 8, 2009, and Anne followed him on May 1, 2009. (Courtesy of Ron Stewart.)

Marjorie (Lucille) Wigton

Born on July 29, 1911, in Aurora, Wigton began instruction on the harp at age 12. She studied in Chicago under the great Italian master Alberto Salvi. She graduated from East Aurora High School in 1929. In 1936, Wiggy became a Zephyrette and acted as a hostess on the Burlington Zephyr trains. She worked for the Burlington Railroad for 39 and a half years, retiring in 1976. Wigton passed away peacefully on October 27, 2011, at age 100. As Victor Hugo has written, "Music expresses that which cannot be said and on which it is impossible to be silent." (Courtesy of Marie and Richard Murphy.)

Harry Clifford Murphy
Harry Murphy was born on August 27, 1892, to Ed and Margaret Murphy in Canton, Illinois. He studied civil and mechanical engineering at Iowa State College, Ames. Murphy moved to Aurora in 1913 and served as an Air Force first lieutenant from 1917 to 1918. He wed Gladys Elizabeth Keating in 1921 in Aurora. Their three children were John, Beth, and Richard. Harry and Gladys settled down at 441 Oak Avenue in 1932. Murphy was selected as president of the Burlington Railroad on August 20, 1949, and would serve until 1965. On September 13, 1965, the Chicago, Burlington's board of directors presented a testimonial that read in part, "a gentleman of the utmost integrity, good humor, quiet strength and modesty, always sensitive to the right and interests of others." On March 7, 1967, a document, Clearance Form A, noted that "Trains will stop at 11 a.m. on March 7 for one minute out of respect for memory of deceased former President Harry C. Murphy." The astute leader of the Burlington Railroad, builder of the first diesel-powered, stainless-steel train in the world in 1934, had passed on March 4, 1967. In 1972, the Fox River Valley Pleasure Driveway and Park District held a dedication of the Harry C. Murphy Miniature Train at Pioneer Park. "Down in the meadow, meadow so low, late in the evening, hear the train blow." (Courtesy of Marie and Richard Murphy.)

CHAPTER NINE

Winsome Wine,
Wonderful in Every Way

*Silently, one by one, in the infinite meadows of heaven
blossomed the lovely stars, the forget-me-nots of the angels.*

—Henry Wadsworth Longfellow

Ruth Armbruster Dieterich Wagner

Elsie Ruth Armbruster Dieterich Wagner was born on Minkler Road in Na-Au-Say Township in Kendall County, on August 7, 1912. Her dad, Herman, was born in 1881 and her mom, Hazel, was born in 1885. Wagner had two brothers, Glenn and Russell, and a sister, Mary Helene. Wagner graduated from East Aurora High School in 1930, and she then worked for Household Finance in Aurora as a business manager. Wagner and her first husband, banker Les Dieterich, raised their son, Thomas Dieterich, at the family home at 1112 Southeast River Road in Montgomery. Tom later graduated from Northwestern and the University of Michigan Law School and became an attorney-at-law. Les and Ruth opened their Fruit Juice House Number 6 business in 1950 at 611 South Broadway, operating it for 15 years. Les died in 1977, and Ruth wed Al Wagner in 1989. Al died in 2007. Ruth's amazing contributions to Aurora include service on the Wayside Cross Board, the Aurora Historical Society, Family Counseling Service, the YWCA, Jennings Terrace, Bardwell School PTA, the United Way, the Kiwanis, and East Aurora Schools. She was named a YWCA Woman of Distinction in 1980, received the Kiwanis's Service to God and Man Award, and was named grand marshal of Montgomery Fest in 2003. She is a member of the Hawthorne Book Club, the DD Chapter of PEO Sisterhood, and the Early American Pressed Glass Club. Wagner has been a member of the First Presbyterian Church in Aurora since 1933. "His master replied, Well done, good and faithful servant." Matthew 25:21.

Sen. Robert Walter Mitchler

Bob Mitchler was born on June 4, 1920, the son of John L. and Clara Rob Mitchler. He graduated from East Aurora High School in 1937. Mitchler wed Helen Drew on June 16, 1950, and they had three children: John, Kurt, and Heidi. Mitchler graduated from Aurora College in 1953, and in 1958, he began working with the Northern Illinois Gas Company. In 1964, he was elected to the Illinois State Senate, where he served for 17 years. Bob was a Republican precinct committeeman for 60 years. He passed peacefully on April 19, 2012, leaving a legacy of community service. He was an advocate of the military and veterans who have served with honor. Bob was a Navy veteran of both World War II and the Korean War.

Warren "Red" Dixon

Dixon was born to Lester and Minnie Dixon on June 22, 1916, in Chicago. He graduated from East Aurora High School. In 1943, he was ordered to active Navy duty and was assigned to the destroyer Escort (DE-438) on May 27, 1944. Returning home, he worked at Commonwealth Edison for 15 years. He and his wife, Daisy, then opened Dixon Realty. Warren was named to the Realtors Hall of Fame, was president of the Aurora Board of Realtors in 1964, and both he and Daisy were Realtor of the Year in 1971. Red coached football at Annunciation School until he was 93 years old. Warren has two children, Mary Agnes and Warren Jr. There are five grandchildren and six great-grandchildren.

The Aurora Roundhouse
The permanent car shops of the Chicago, Burlington, & Quincy Railroad were constructed in 1855–1856 at a cost of $150,000. It is the oldest full roundhouse in America and the only one remaining. The National Park Service has deemed the complex "of preeminent importance in American transportation history." The building is listed in the National Register of Historic Places. The Chicago, Burlington, & Quincy repair facility closed in 1974, and a portion was demolished in 1977. City officials then formulated plans to create an intermodal transportation complex in 1988. The Walter Payton Roundhouse Restaurant operated at the south end from 1996 until 2011, when Two Brothers assumed ownership of the restaurant and brewery complex. (Author collection.)

Art Stiegleiter
Growing up at 520 South Broadway in Aurora, Stiegleiter was the son of Ruth May and George F. Stiegleiter. He was born on January 22, 1918. In 1937, he married Ruth Nichols, and they had two children, Arthur Jr. and Sandra. Ruth passed on January 16, 1981. Stiegleiter married Lois Schoen on November 13, 1981, and the couple had 25 years together before her death in 2006. Stiegleiter has two grandsons, Jason and Brian. He remembers checking Mayor Egan's office for "bugs" but finding no recording devices. He became an Aurora fireman in 1952 at a salary of $6,000. Stiegleiter was a captain before retiring in 1976. (Courtesy Aurora Regional Fire Museum archives.)

Minerva Coterie Literary Society

In the horse-and-buggy days, most women did not pursue higher education. A group of 13 Aurora women founded the Minerva Coterie in 1882 as "a society organized for the purpose of mutual improvement in English literature, art, science and social customs." They chose the name for the group from the Roman goddess of wisdom. Today, 130 years later, the 34 members meet twice a month from October until May to enjoy the erudite discussions, the camaraderie, and the refreshments. Shown in the photograph are the members in 1952 as the group celebrated its 70th year.

Gladys Larson Mason

As this book went to press, Mason was the oldest living resident in the city of Aurora, at age 106. She was born on March 14, 1906, to Hazel and Frank Larson. She lived with her grandmother Dahlia Boosie Risely in 1924 while she found a job at the Aurora Corset Factory. She wed Albert Mason on September 4, 1926, and they had two children, Jimmy and Betty. She has two grandchildren, Deborah and Jeff, and two great-grandchildren, Matthew and Joseph. Mason was a founding member in 1946 of the Fox Valley Animal Welfare League. On March 2, 2012, Mayor Thomas J. Weisner proclaimed her to be Aurora's oldest living resident and visited Mason and her daughter, Betty Reicher, to present the award on March 19, 2012. Gladys died on August 26, 2012.

The Aurora Farmers Market

On April 3, 1912, C.B. Coleman, a farmer from Warrenville, sold his 30-dozen eggs at 20¢ a dozen and a barrel of apples at the first day of the Aurora market. The 2012 main market featured 50 vendors. Locals sold produce, jewelry, and farm fresh items. The Wiltse's Farm Produce family, and the Theis family, both from Maple Park, have been coming for the past century. The photograph shows the Theis Farm stand.

Marion Breese Glass Carter

The Breese family were early settlers in Aurora. Maternal grandmother Nell Heath was a ticket seller at the Paramount Theatre. Three of her sons, Bill, Tom, and Steven Glass, served with the Aurora Fire Department. Marion was born on June 23, 1907, to William Lewis Breese and Nell Elizabeth Heath. She graduated from St. Mary's Grade School and East Aurora High School, and went to work at age 16 for Illinois Bell. Marion and Walter Thomas Glass wed on June 6, 1926, and they had 10 children. She died at the age of 90 on September 1, 1997, but not before laboriously compiling her family history, tracing it back 37 generations.

Phillips Park

The deed for the park was signed on November 21, 1899, and records indicate that as early as 1835, the hilly area with wooded coves and lagoons was used for picnics and family outings. In 1899, city park was purchased by the estate of Travis Phillips, former grocer, alderman, and mayor. He instructed that $24,000 from his estate be used to purchase property for a park, and donated it to the city. A total of 60 acres, at $400 per acre, were purchased and turned over. The park was renamed Phillips Park in 1902. The zoo was established in 1915. On March 7, 1934, prehistoric Mastodon bones and tusks between 10,000 and 22,000 years old were discovered in the park. The new Phillips Park Visitors Center and Mastodon Gallery, costing $650,000, opened to the public in the autumn of 2003. It was renamed the David and Karen Stover Visitor Center on May 22, 2008.

The Luxemburger Club

On June 16, 1890, the Luxemburger Club was founded. The first Luxemburgers to settle in Aurora came during 1850–1860. On June 17, 1917, the new hall was dedicated. At this time, the English language was spoken at meetings instead of the Luxemburger tongue. (Photograph restoration by Dave Chrestenson.)

Irene Campbell

In 1946, the Fox Valley Animal Welfare League sponsored a pet show at West High School. Mrs. Myron W. Larson was chair of the pet show. Today, 66 years later, the Fox Valley Animal Welfare League continues to rescue animals. Irene Campbell, age 93, a volunteer since the 1950s and group treasurer since 1992, is shown in this 1926 photograph.

Al DeSotell

Born at home at 804 Spring Street in Aurora on October 13, 1916, Al DeSotell's parents were Joseph "Mose" DeSotell and Azilda Regnier DeSotell. Al's grandfather, Isaac Desautels, had fought in the Civil War. DeSotell worked in the sheet-metal business with his father before serving in the Army in World War II. On June 28, 1952, he wed Abbie Reuter, and the couple had two children, Kathryn and Mary Ann. DeSotell worked for the CB&Q Railroad, retiring in 1979. On March 2, 2012, Mayor Thomas J. Weisner presented DeSotell with a framed certificate during Aurora's 175th birthday celebration held at the Aurora Roundhouse. (Courtesy of Katie DeSotell.)

Riverfront Playhouse

In 1978, David Morris created the Riverfront Playhouse in a small area of Max's All-American Bar on Galena Boulevard. It later relocated to 11-13 Water Street Mall. *Under The Yum Yum Tree* was the first production, and there have been 150 shows since that opening. All of the furniture and costumes were destroyed in the 1996 area flood, but most inventory pieces have since been replaced. The Riverfront Playhouse continues, 35 years after its founding, as a diverse and exciting entertainment venue. (Author's collection.)

The Blanford Clock

In 1905, the Blanford Astronomical Clock was described as a "universe in miniature" because of its sophisticated mechanisms. The clock, with a hand-carved mahogany case, gives the local time, the date, phases of the moon, signs of the zodiac, and times of principal cities around the world. William Blanford was born in London, England, in 1838 to a family of expert clockmakers. He came to the United States in his forties and worked at the Elgin National Watch Company and the Aurora Watch Company. He spent five years building the clock in a workshop next to his home on Douglas Avenue. The clock is nine feet tall, three feet wide, and two feet deep. The mechanisms include a stamped plate of corresponding longitudinal degrees to calculate time in other cities. The city purchased the clock for $5,000 upon his death on February 18, 1920. Since 1944, it has been in the permanent collection of the Aurora Historical Society. (Author's collection.)

BIBLIOGRAPHY

Art and Society. Central Illinois Business Publishers, January/February 2012.

"AU makes multimillion dollar contribution to region." *Suburban News Bulletin*. February 15, 2012.

Aurora. Greater Aurora Chamber of Commerce. Profile publication. 1980.

Aurora Borealis, newsletter. City of Aurora. Winter 2011.

Aurora Business Journal. Vol. 1, Issue 17. December 1, 1988.

Aurora, Illinois Illustrated. 1890. Rathbone, Sard & Co., March 13, 1890.

Aurora Life magazine, 1990–1991.

Aurora Police Department 2010. Compiled by the Police Benevolent & Protective Association.

Barclay, Robert. *Aurora 1837–1987*. Aurora, IL: Copley Press, 1988.

Barclay, Robert W., and Vernon Derry. *Aurora, the City of Bridges*. Aurora, IL: Aurora Historical Society, 1955.

Buck, Dennis. *From Slavery to Glory*. Aurora, IL: Aurora Historical Society, 2005.

Building Aurora. Sears Houses in Aurora, Illinois. Aurora, IL: Aurora Preservation Commission.

Burton, Charles Pierce. *Aurora: From covered wagon to stream-lined zephyr*. 1937.

"Cat and Mouse are Pals." *Chicago Daily Tribune*, May 26, 1946, p. W4.

Civil War Era Architecture of Aurora. Preservation Commission. 2008.

Derry, Vernon. *Aurora . . . in the beginning*. Aurora, 1953.

Distinguished Alumni EAHS Awards Ceremony program. April 28, 2011.

Downtown Auroran magazine. Spring/Summer, 2010. Marissa Amoni, ed.

Ellis, Edward S. *The Eclectic Primary History of the United States*. Cincinnati: Van Antwerp, Bragg & Co., 1884.

"Fox Symphony Readies 11th Season." *Chicago Tribune*, October 17, 1968.

Greenaway, J.W. *With the Colors from Aurora, 1917–1919*. Aurora, IL: J.W. Greenaway, 1919.

Handford, Thomas W. *The Sands of Time. A Book of Birthday Gems*. Chicago: Donohue, Henneberry, and Co., 1885.

Higgins, Jo Fredell. *Aurora: A Postcard History*. Charleston, SC: Arcadia Publishing, 2006.

Hopp, Nancy S. *A History of the Sisters of Mercy in Aurora, Ill. 1910–1992*.

Hunton, Daniel T.V. *Philip Hunton and his Descendants*. Canton, MA: Cambridge University Press, 1881.

Illinois Women. "75 Years of the Right to Vote." Chicago, IL: *Chicago Sun-Times* Features. 1996.

Joslyn, R. Waite, and Frank W. Joslyn. *History of Kane County, Ill*. Chicago, 1908.

Journal of Proceedings and Addresses of the 43 Annual Meeting of the NEA. June 1904.

"Lauzen's run a test for Illinois GOP." *Tri-Cities Daily Herald*, March 5, 2012, p. 1.

"Liberty's Victorious Conflict." Chicago: The Magazine Circulation Co. Inc., 1918.

Life Lines newsletter. Fox Valley Animal Welfare League. July 2011.

MEDC 2006 Annual Dinner Reception and Longevity Awards. February 9, 2006.

Medernach, John S. *Banking at the Corner of River and Downer: The History of the Old Second National Bank*. Aurora, IL: Second National Bank, 2000.

"Oh, for the Wail of That Whistle!" *Beacon News* Burlington Centennial Edition, May 19, 1964.

Ormond, Mary Clark. *In and Through Our Gardens*. The Tuesday Garden Club of Aurora. 1936–2011.

Pages of Time. 1938. Goodlettsville, TN.

Peoria Progress magazine, Central Illinois Business Publishers, 2012.

"Railroad Head Just a Friend to Employees," *Beacon News Burlington* Centennial Edition. May 19, 1964.

Republican News. Robert Sauceda, ed. Volume 1. Paid advertisement. March 2012.

St. John's United Church of Christ. Shutterfly online pictorial of Mary's Table, 2009.

Schaffer, Karen A. *The Maud Powell Society for Music and Education Souvenir*. Winter 2010–2011.

Vogel, Hy. "Prominent Men of Aurora and the Fox Valley." *Beacon-News*, 1935.

INDEX

Find more books like this at
www.legendarylocals.com

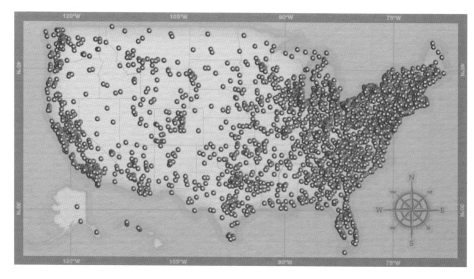

Discover more local and regional history books at
www.arcadiapublishing.com